FM 3-25.26

MAP READING
AND
LAND NAVIGATION

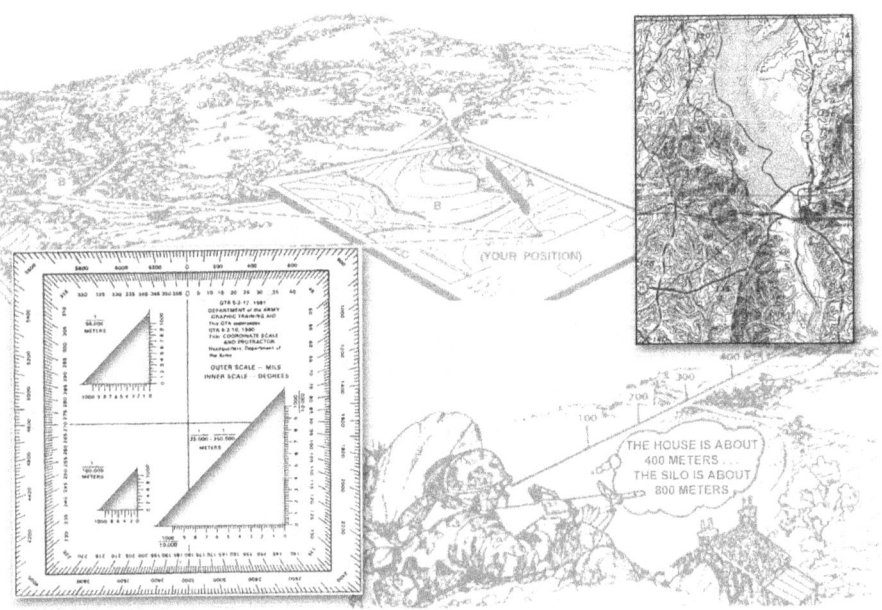

January 2005

HEADQUARTERS
DEPARTMENT OF THE ARMY

DISTRIBUTION RESTRICTION: Distribution authorized to U.S. Government agencies and their contractors only (this publication addresses current technology in areas of significant or potentially significant military application). This determination was made on September 13, 2004. Other requests for this document will be referred to: Commandant; U.S. Army Infantry School; ATTN: ATSH-INB-O; Fort Benning, GA 31905-5593.

DESTRUCTION NOTICE: Destroy by any method that will prevent disclosure of contents or reconstruction of the document.

FOUO

This publication is available at
Army Knowledge Online (www.us.army.mil) and
General Dennis J. Reimer Training and Doctrine
Digital Library at (http://www.train.army.mil)

FM 3-25.26
C1

Change 1

Headquarters
Department of the Army
Washington, DC, 30 August 2006

Map Reading and Land Navigation

1. Change FM 3-25.26, 18 January 2005, as follows:

Remove old pages:	**Insert new pages:**
Preface v through vi	Preface v through vi
1-1 through 1-2	1-1 through 1-2
4-7 through 4-8	4-7 through 4-8
4-19 through 4-20	4-19 through 4-20
Glossary 1 through 4	Glossary 1 through 2

2. A star (*) marks new or changed material. A star (*) on the revised contents page could indicate a minor (single word) or entire paragraph change.

3. File this transmittal sheet in front of the publication.

By Order of the Secretary of the Army:

PETER J. SCHOOMAKER
General, United States Army
Chief of Staff

Official:

JOYCE E. MORROW
Administrative Assistant to the
Secretary of the Army

DISTRIBUTION RESTRICTION: Distribution authorized to U.S. Government agencies and their contractors only (this publication addresses current technology in areas of significant or potentially significant military application). This determination was made on September 13, 2004. Other requests for this document will be referred to: Commandant; U.S. Army Infantry School; ATTN: ATSH-INB-O; Fort Benning, GA 31905-5593.

DESTRUCTION NOTICE: Destroy by any method that will prevent disclosure of contents or reconstruction of the document.

DISTRIBUTION: Regular Army, Army National Guard, and U.S. Army Reserve: To be distributed in accordance with initial distribution number 110166 requirements for FM 3-25.26.

FOUO

This Page intentionally left blank.

FIELD MANUAL
No. 3-25.26

*FM 3-25.26

HEADQUARTERS
DEPARTMENT OF THE ARMY
Washington, DC, 18 January 2005

MAP READING AND LAND NAVIGATION

CONTENTS

Page

PREFACE v

Part One
MAP READING

CHAPTER 1. TRAINING STRATEGY
- 1-1. Building-Block Approach 1-1
- 1-2. Armywide Implementation 1-2
- 1-3. Safety 1-2

CHAPTER 2. MAPS
- 2-1. Definition 2-1
- 2-2. Purpose 2-1
- 2-3. Procurement 2-2
- 2-4. Security 2-2
- 2-5. Care 2-3
- 2-6. Categories 2-3
- 2-7. Military Map Substitutes 2-6
- 2-8. Standards of Accuracy 2-7

CHAPTER 3. MARGINAL INFORMATION AND SYMBOLS
- 3-1. Marginal Information on a Military Map 3-1
- 3-2. Additional Notes 3-5
- 3-3. Topographic Map Symbols 3-5
- 3-4. Military Symbols 3-6
- 3-5. Colors Used on a Military Map 3-6

DISTRIBUTION RESTRICTION: Distribution authorized to U.S. Government agencies and their contractors only (this publication addresses current technology in areas of significant or potentially significant military application). This determination was made on September 13, 2004. Other requests for this document will be referred to: Commandant; U.S. Army Infantry School; ATTN: ATSH-INB-O; Fort Benning, GA 31905-5593.

DESTRUCTION NOTICE: Destroy by any method that will prevent disclosure of contents or reconstruction of the document.

*This publication supersedes FM 3-25.26, 20 July 2001.

FM 3-25.26

                              Page

CHAPTER 4. GRIDS
- 4-1. Reference System 4-1
- 4-2. Geographic Coordinates 4-1
- 4-3. Military Grids 4-10
- 4-4. United States Army Military Grid Reference System 4-12
- 4-5. Locating a Point Using Grid Coordinates 4-17
- 4-6. Locating a Point Using the U.S. Army Military Grid Reference System 4-19
- 4-7. Grid Reference Box 4-22
- 4-8. Other Grid Systems 4-23
- 4-9. Protection of Map Coordinates and Locations 4-25

CHAPTER 5. SCALE AND DISTANCE
- 5-1. Representative Fraction 5-1
- 5-2. Graphic (Bar) Scales 5-3
- 5-3. Other Methods 5-10

CHAPTER 6. DIRECTION
- 6-1. Methods of Expressing Direction 6-1
- 6-2. Base Lines 6-1
- 6-3. Azimuths 6-2
- 6-4. Grid Azimuths 6-3
- 6-5. Protractor 6-4
- 6-6. Declination Diagram 6-7
- 6-7. Intersection 6-13
- 6-8. Resection 6-15
- 6-9. Modified Resection 6-16
- 6-10. Polar Plot 6-17

CHAPTER 7. OVERLAYS
- 7-1. Purpose 7-1
- 7-2. Map Overlay 7-1
- 7-3. Aerial Photograph Overlay 7-3

CHAPTER 8. AERIAL PHOTOGRAPHS
- 8-1. Comparison with Maps 8-1
- 8-2. Types 8-1
- 8-3. Types of Film 8-7
- 8-4. Numbering and Titling Information 8-7
- 8-5. Scale Determination 8-8
- 8-6. Indexing 8-10
- 8-7. Orientation of Photograph 8-13
- 8-8. Point Designation Grid 8-14

		Page
8-9.	Identification of Photograph Features	8-17
8-10.	Stereovision	8-18

Part Two
LAND NAVIGATION

CHAPTER 9. NAVIGATION EQUIPMENT AND METHODS
9-1.	Types of Compasses	9-1
9-2.	Lensatic Compass	9-1
9-3.	Compass Handling	9-2
9-4.	Using a Compass	9-3
9-5.	Field-Expedient Methods	9-7
9-6.	Global Positioning System	9-12

CHAPTER 10. ELEVATION AND RELIEF
10-1.	Definitions	10-1
10-2.	Methods of Depicting Relief	10-1
10-3.	Contour Intervals	10-2
10-4.	Types of Slopes	10-5
10-5.	Percentage of Slope	10-7
10-6.	Terrain Features	10-10
10-7.	Interpretation of Terrain Features	10-16
10-8.	Profiles	10-19

CHAPTER 11. TERRAIN ASSOCIATION
11-1.	Orientation of the Map	11-1
11-2.	Locations	11-6
11-3.	Terrain Association Usage	11-6
11-4.	Tactical Considerations	11-9
11-5.	Movement and Route Selection	11-12
11-6.	Navigation Methods	11-14
11-7.	Night Navigation	11-18

CHAPTER 12. MOUNTED LAND NAVIGATION
12-1.	Principles	12-1
12-2.	Navigator's Duties	12-1
12-3.	Movement	12-1
12-4.	Terrain Association Navigation	12-3
12-5.	Dead Reckoning Navigation	12-5
12-6.	Stabilized Turret Alignment Navigation	12-7
12-7.	Combination Navigation	12-7

	Page
CHAPTER 13. NAVIGATION IN DIFFERENT TYPES OF TERRAIN	
13-1. Desert Terrain	13-1
13-2. Mountain Terrain	13-4
13-3. Jungle Terrain	13-6
13-4. Arctic Terrain	13-9
13-5. Urban Areas	13-10
CHAPTER 14. UNIT SUSTAINMENT	
14-1. Set Up a Sustainment Program	14-1
14-2. Set Up a Train-the-Trainer Program	14-2
14-3. Set Up a Land Navigation Course	14-2
APPENDIX A. SKETCHES	A-1
APPENDIX B. MAP FOLDING TECHNIQUES	B-1
APPENDIX C. UNITS OF MEASURE AND CONVERSION FACTORS	C-1
APPENDIX D. JOINT OPERATIONS GRAPHICS	D-1
APPENDIX E. ORIENTEERING	E-1
APPENDIX F. M2 COMPASS	F-1
APPENDIX G. ADDITIONAL AIDS	G-1
APPENDIX H. FOREIGN MAPS	H-1
APPENDIX I. GLOBAL POSITIONING SYSTEM	I-1
APPENDIX J. PRECISION LIGHTWEIGHT GPS RECEIVER	J-1
APPENDIX K. DEFENSE ADVANCED GPS RECEIVER	K-1
GLOSSARY	Glossary-1
REFERENCES	References-1
INDEX	Index-1

Change 1, FM 3-25.26

PREFACE

The field manual provides a standardized source document for Armywide reference on *map reading* and *land navigation*. It applies to every Soldier in the Army regardless of service branch, MOS, or rank.

* This manual also contains both doctrine and training guidance on map reading and land navigation. Part One addresses map reading and Part Two, land navigation. The appendixes include an introduction to orienteering and a discussion of several devices that can assist the Soldier in land navigation.

* The proponent for this publication is the U.S. Army Training and Doctrine Command. The preparing agency is the U.S. Army Infantry School. You may send comments and recommendations by any means, US mail, e-mail, fax, or telephone, as long as you use or follow the format of DA Form 2028, *Recommended Changes to Publications and Blank Forms*. You may also phone for more information. Point of contact information is as follows.

    E-mail: 229-DOC-LIT@benning.army.mil
    Phone: Commercial: 706-545-8623 or DSN: 835-8623
    Fax: Commercial: 706-545-8600 or DSN: 835-8600
    US Mail: Commandant, USAIS
             ATTN; ATSH-INB, BLDG 74, Room 102
             Dilboy Street, Bldg 74, Rm 102
             Fort Benning, GA 31905-5593

Unless this publication states otherwise, masculine nouns and pronouns do not refer exclusively to men.

FOUO

This Page intentionally left blank.

PART ONE
MAP READING

CHAPTER 1
TRAINING STRATEGY

This manual responds to an Armywide need for a new map reading and land navigation training strategy based on updated doctrine. This chapter describes and illustrates this approach to teaching these skills.

1-1. BUILDING-BLOCK APPROACH

Institution courses are designed to prepare the Soldier for a more advanced duty position in his unit. The critical soldiering skills of move, shoot, and communicate must be trained, practiced, and sustained at every level in the schools as well as in the unit. The map reading and land navigation skills taught at each level are critical to the soldiering skills of the duty position for which he is being school-trained. Therefore, they are also a prerequisite for a critical skill at a more advanced level.

    a. A Soldier completing initial-entry training must be prepared to become a team member. He must be proficient in the basic map reading and dead reckoning skills.

*   b. After completing the Warrior Leader Course (WLC), a Soldier should be ready to be a team leader. This duty position requires expertise in the skills of map reading, dead reckoning, and terrain association.

    c. A Soldier completing the Basic Noncommissioned Officer's Course (BNCOC) has been trained for the squad leader position. Map reading and land navigation at skill level 3 requires development of problem-solving skills; for example, route selection and squad tactical movement.

    d. At skill level 4, the Soldier completing the Advanced Noncommissioned Officer's Course (ANCOC) is prepared to assume the duty position of platoon sergeant or operations NCO. Planning tactical movements, developing unit sustainment, and making decisions are the important land navigation skills at this level.

    e. Officers follow similar progression. A new second lieutenant must have mastered map reading and land navigation skills, and have an aptitude for dead reckoning and terrain association.

    (1) After completing the Officer Basic Course (OBC), the officer must be prepared to assume the duties and responsibilities of a platoon leader. He is required to execute the orders and operations of his commander. Map reading and land navigation at this level require development of the problem-solving skills of route selection and tactical movement.

*(2) After completing the Captain's Career Course (CCC), the officer is prepared to assume the duties and responsibilities of a company commander or primary staff officer. The commander must plan and execute operations with full consideration to all aspects of navigation. The staff officer must recommend battlefield placement of all administrative, logistical, and personnel resources. These recommendations cannot be tactically sound unless the estimate process includes a detailed analysis of the area of operations. This ability requires expertise in all map reading and navigation skills to include the use of nonmilitary maps, aerial photographs, and terrain analysis with respect to both friendly and enemy forces. The commander/staff officer must plan and execute a program to develop the unit's train-the-trainer program for land navigation.

30 August 2006

FOUO

f. A program of demonstrated proficiency of all the preceding skill levels to the specified conditions and standards is a prerequisite to the successful implementation of a building-block training approach. This approach reflects duty position responsibilities in map reading and land navigation. An understanding of the fundamental techniques of dead reckoning or field-expedient methods is a basic survival skill that each Soldier must develop at the initial-entry level. This skill provides a support foundation for more interpretive analysis at intermediate skill levels 2 and 3, with final progression to level 4. Mastery of all map reading and land navigation tasks required in previous duty positions is essential for the sequential development of increasingly difficult abilities. Scope statements support the building-block approach. It is part of the training doctrine at each level in the institutional training environment of each course.

*

1-2. ARMYWIDE IMPLEMENTATION

A mandatory core of critical map reading and land navigation tasks and a list of electives will be provided to each TRADOC service school and FORSCOM professional development school. Standardization is achieved through the mandatory core.

1-3. SAFETY

Unit leaders plan to brief and enforce all safety regulations established by local range control. They coordinate the mode of evacuation of casualties through the appropriate channels. They review all installation safety regulations. Unit leaders must complete a thorough terrain reconnaissance before using an area for land navigation training. They should look for dangerous terrain, heavily trafficked roads, water obstacles, wildlife, and training debris.

CHAPTER 2
MAPS

Cartography is the art and science of expressing the known physical features of the earth graphically by maps and charts. No one knows who drew, molded, laced together, or scratched out in the dirt the first map. But a study of history reveals that the most pressing demands for accuracy and detail in mapping have come as the result of military needs. Today, the complexities of tactical operations and deployment of troops are such that it is essential for all Soldiers to be able to read and interpret their maps in order to move quickly and effectively on the battlefield. This chapter includes the definition and purpose of a map and describes map security, types, categories, and scales.

2-1. DEFINITION

A map is a graphic representation of a portion of the earth's surface drawn to scale, as seen from above. It uses colors, symbols, and labels to represent features found on the ground. The ideal representation would be realized if every feature of the area being mapped could be shown in true shape. Obviously this is impossible, and an attempt to plot each feature true-to-scale would result in a product impossible to read even with the aid of a magnifying glass.

   a. To be understandable, features must be represented by conventional signs and symbols. To be legible, many of these must be exaggerated in size, often far beyond the actual ground limits of the feature represented. On a 1:250,000-scale map, the prescribed symbol for a building covers an area about 500 square feet on the ground; a road symbol is equivalent to a road about 520 feet wide on the ground; the symbol for a single-track railroad (the length of a cross-tie) is equivalent to a railroad cross-tie about 1,000 feet on the ground.

   b. The portrayal of many features requires similar exaggeration. Therefore, the selection of features to be shown, as well as their portrayal, is in accordance with the guidance established by the National Geospatial-Intelligence Agency (NGA).

2-2. PURPOSE

A map provides information on the existence of, the location of, and the distance between ground features such as populated places and routes of travel and communication. It also indicates variations in terrain, heights of natural features, and the extent of vegetation cover. With our military forces dispersed throughout the world, it is necessary to rely on maps to provide information to our combat elements and to resolve logistical operations far from our shores. Soldiers and materials must be transported, stored, and placed into operation at the proper time and place. Much of this planning must be done using maps. All operations require a supply of maps; however, the finest maps available are worthless unless the map user knows how to read them.

2-3. PROCUREMENT

Most military units are authorized a basic load of maps. Local command supplements to AR 115-11 provide tables of initial allowances for maps. Map requisitions and distributions are accomplished through the NGA Hydrographic and Topographic Center's Office of Distribution and Services. In the division, however, the G2 section is responsible for maps.

   a. To order a map, refer to the NGA catalog located at your S2/G2 shop. Part 3 of this catalog, Topographic Maps, has five volumes. Using the delineated map index, find the map or maps you want based upon the location of the nearest city. With this information, order maps using the following forms:

- **Department of Defense Form 1348 (DOD Single Line Item Requisition System Document [Manual]).** You can order copies of only one map sheet on each form.
- **Department of Defense Form 1348M (DOD Single Line Item Requisition System Document [Mechanical]).** DD 1348M is a punch card form for AUDODIN ordering.

The numbered sections of all forms are the same. For example: In block 1, if you are in CONUS, enter "AOD;" if you are overseas, enter "AO4." In block 2, use one of the following codes for your location. Your supply section will help you complete the rest of the form.

LOCATION	CODE
Europe	CS7
Hawaii	HM9
Korea	WM4
Alaska	WC1
Panama	HMJ
CONUS	HM8

   b. Stock numbers are also listed in map catalogs, which are available at division and higher levels and occasionally in smaller units. A map catalog consists of small-scale maps upon which the outlines of the individual map sheets of a series have been delineated. Another document that is an aid to the map user is the gazetteer. A gazetteer lists all the names appearing on a map series of a geographical area, a designation that identifies what is located at that place name, a grid reference, a sheet number of the map upon which the name appeared, and the latitude and longitude of the named features. Gazetteers are prepared for maps of foreign areas only.

2-4. SECURITY

All maps should be considered as documents that require special handling. If a map falls into unauthorized hands, it could easily endanger military operations by providing information of friendly plans or areas of interest to the enemy. Even more important would be a map on which the movements or positions of friendly Soldiers were marked. It is possible, even though the markings on a map have been erased, to determine some of the erased information. *Maps are documents that must not fall into unauthorized hands.*

   a. If a map is no longer needed, it must be turned in to the proper authority. If a map is in danger of being captured, it must be destroyed. The best method of destruction is by burning it and scattering the ashes. If burning is not possible, the map can be torn into small pieces and scattered over a wide area.

b. Maps of some areas of the world are subject to third party limitations. These are agreements that permit the United States to make and use maps of another country provided these maps are not released to any third party without permission of the country concerned. Such maps require special handling.

c. Some maps may be classified and must be handled and cared for in accordance with AR 380-5 and, if applicable, other local security directives.

2-5. CARE

Maps are documents printed on paper and require protection from water, mud, and tearing. Whenever possible, a map should be carried in a waterproof case, in a pocket, or in some other place where it is handy for use but still protected. (Appendix B shows two ways of folding a map.)

a. Care must also be taken when using a map since it may have to last a long time. A pencil is recommended if marking a map becomes necessary. Use light lines so they may be erased easily without smearing and smudging or leaving marks that may cause confusion later. If the map margins must be trimmed for any reason, it is essential to note any marginal information that may be needed later, such as grid data and magnetic declination.

b. Special care should be taken of a map that is being used in a tactical mission, especially in small units; the mission may depend on that map. All members of such units should know the map's location at all times.

2-6. CATEGORIES

The NGA's mission is to provide mapping, charting, and all geodesy support to the armed forces and all other national security operations. NGA produces four categories of products and services: hydrographic, topographic, aeronautical, and missile and targeting. Military maps are categorized by scale and type.

a. **Scale.** Because a map is a graphic representation of a portion of the earth's surface drawn to scale as seen from above, it is important to know what mathematical scale has been used. You must know the scale to determine ground distances between objects or locations on the map, the size of the area covered, and how the scale may affect the amount of detail being shown. The mathematical scale of a map is the ratio or fraction between the distance on a map and the corresponding distance on the surface of the earth. Scale is reported as a representative fraction (RF) with the map distance as the numerator and the ground distance as the denominator.

$$\text{Representative fraction (scale)} = \frac{\text{map distance}}{\text{ground distance}}$$

As the denominator of the representative fraction gets larger and the ratio gets smaller, the scale of the map decreases. NGA maps are classified by scale into three categories: small-, medium-, and large-scale maps (Figure 2-1, page 2-4). The terms *"small scale," "medium scale,"* and *"large scale"* may be confusing when read in conjunction with the number. However, if the number is viewed as a fraction, it quickly becomes apparent that 1:600,000

FM 3-25.26

of something is smaller than 1:75,000 of the same thing. Therefore, the larger the number after 1:, the smaller the scale of the map.

(1) *Small*. Maps with scales of 1:1,000,000 and smaller are used for general planning and for strategic studies (bottom map in Figure 2-1). The standard small-scale map is 1:1,000,000. This map covers a very large land area at the expense of detail.

(2) *Medium*. Maps with scales larger than 1:1,000,000 but smaller than 1:75,000 are used for operational planning (center map in Figure 2-1). They contain a moderate amount of detail, but terrain analysis is best done with the large-scale maps. The standard medium-scale map is 1:250,000. Medium-scale maps of 1:100,000 are also frequently encountered.

(3) *Large*. Maps with scales of 1:75,000 and larger are used for tactical, administrative, and logistical planning (top map in Figure 2-1). These are the maps that you as a Soldier or junior leader are most likely to encounter. The standard large-scale map is 1:50,000; however, many areas have been mapped at a scale of 1:25,000.

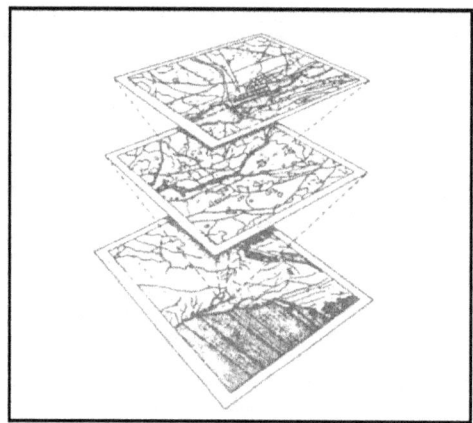

Figure 2-1. Scale classifications.

b. **Types**. The map of choice for land navigators is the 1:50,000-scale military topographic map. It is important, however, to know how to use the many other products available from the NGA as well. When operating in foreign places, you may discover that NGA map products have not yet been produced to cover your particular area of operations, or they may not be available to your unit when you require them. Therefore, you must be prepared to use maps produced by foreign governments that may or may not meet the standards for accuracy set by NGA. These maps often use symbols that resemble those found on NGA maps but which have completely different meanings. There may be other times when you must operate with any map you can obtain. This might be a commercially produced map run off on a copy machine at higher headquarters. In Grenada, many of our troops used a British tourist map.

(1) *Planimetric Map*. A planimetric map presents only the horizontal positions for the features represented. It is distinguished from a topographic map by the omission of relief, normally represented by contour lines. Sometimes, it is called a line map.

FOUO

18 January 2005

(2) ***Topographic Map***. A topographic map portrays terrain features in a measurable way, as well as the horizontal positions of the features represented. The vertical positions, or relief, are normally represented by contour lines on military topographic maps. On maps showing relief, the elevations and contours are measured from a specific vertical datum plane, usually mean sea level. (Figure 3-1 in Chapter 3 shows a typical topographic map.)

(3) ***Photomap***. A photomap is a reproduction of an aerial photograph upon which grid lines, marginal data, place names, route numbers, important elevations, boundaries, and approximate scale and direction have been added. (See Chapter 8 for additional information on aerial photographs.)

(4) ***Joint Operations Graphics***. Joint operations graphics are based on the format of standard 1:250,000 medium-scale military topographic maps, but they contain additional information needed in joint air-ground operations (Figure 2-2). Along the north and east edges of the graphic, detail is extended beyond the standard map sheet to provide overlap with adjacent sheets. These maps are produced both in ground and air formats. Each version is identified in the lower margin as either joint operations graphic (air) or joint operations graphic (ground). The topographic information is identical on both, but the ground version shows elevations and contour in meters and the air version shows them in feet. Layer (elevation) tinting and relief shading are added as an aid to interpolating relief. Both versions emphasize airlanding facilities (shown in purple), but the air version has additional symbols to identify aids and obstructions to air navigation. (See Appendix D for additional information.)

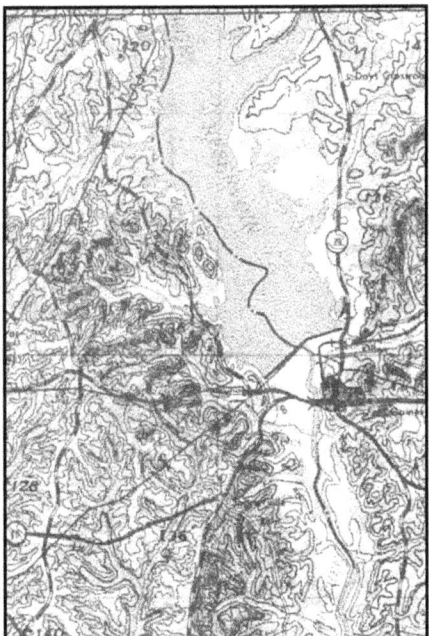

Figure 2-2. Joint operations graphic (air).

(5) **Photomosaic**. A photomosaic is an assembly of aerial photographs that is commonly called a mosaic in topographic usage. Mosaics are useful when time does not permit the compilation of a more accurate map. The accuracy of a mosaic depends on the method employed in its preparation and may vary from simply a good pictorial effect of the ground to that of a planimetric map.

(6) **Terrain Model**. A terrain model is a scale model of the terrain showing features, and in large-scale models showing industrial and cultural shapes. It provides a means for visualizing the terrain for planning or indoctrination purposes and for briefing on assault landings.

(7) **Military City Map**. A military city map is a topographic map (usually at 1:12,550 scale, sometimes up to 1:5,000), showing the details of a city. It delineates streets and shows street names, important buildings, and other elements of the urban landscape important to navigation and military operations in urban terrain. The scale of a military city map depends on the importance and size of the city, density of detail, and available intelligence information.

(8) **Special Maps**. Special maps are for special purposes such as trafficability, communications, and assault maps. They are usually in the form of an overprint in the scales smaller than 1:100,000 but larger than 1:1,000,000. A special purpose map is one that has been designed or modified to give information not covered on a standard map. The wide range of subjects that could be covered under the heading of special purpose maps prohibits, within the scope of this manual, more than a brief mention of a few important ones. Some of the subjects covered are—

- Terrain features.
- Drainage characteristics.
- Vegetation.
- Climate.
- Coasts and landing beaches.
- Roads and bridges.
- Railroads.
- Airfields.
- Urban areas.
- Electric power.
- Fuels.
- Surface water resources.
- Ground water resources.
- Natural construction materials.
- Cross-country movements.
- Suitability for airfield construction.
- Airborne operations.

2-7. MILITARY MAP SUBSTITUTES

If military maps are not available, use substitute maps. The substitute maps can range from foreign military or commercial maps to field sketches. The NGA can provide black and white reproductions of many foreign maps and can produce its own maps based upon intelligence.

a. **Foreign Maps**. Foreign maps have been compiled by nations other than our own. When they must be used, the marginal information and grids are changed to conform to our standards, if time permits. The scales may differ from our maps, but they do express the ratio of map distance to ground distance and can be used in the same way. The legend must be used since the map symbols almost always differ from ours. Because the accuracy of foreign maps varies considerably, they are usually evaluated in regard to established accuracy standards before they are issued to our troops. (See Appendix H for additional information.)

b. **Atlases**. Atlases are collections of maps of regions, countries, continents, or the world. Such maps are accurate only to a degree and can be used for general information only.

c. **Geographic Maps**. Geographic maps provide an overall idea of the mapped area in relation to climate, population, relief, vegetation, and hydrography. They also show the general location of major urban areas.

d. **Tourist Road Maps**. Tourist road maps are maps of a region in which the main means of transportation and areas of interest are shown. Some of these maps show secondary networks of roads, historic sites, museums, and beaches in detail. They may contain road and time distance between points. The scale should be carefully considered when using these maps.

e. **City/Utility Maps**. City/utility maps are maps of urban areas showing streets, water ducts, electricity and telephone lines, and sewers.

f. **Field Sketches**. Field sketches are preliminary drawings of an area or piece of terrain. (See Appendix A.)

g. **Aerial Photographs**. Aerial photographs can be used as map supplements or substitutes to help you analyze the terrain, plan your route, or guide your movement. (See Chapter 8 for additional information).

2-8. STANDARDS OF ACCURACY

Accuracy is the degree of conformity with which horizontal positions and vertical values are represented on a map in relation to an established standard. This standard is determined by the NGA based on user requirements. Maps are considered to meet accuracy requirement standards unless otherwise specified in the marginal information.

This Page intentionally left blank.

FM 3-25.26

CHAPTER 3
MARGINAL INFORMATION AND SYMBOLS

A map could be compared to any piece of equipment, in that before it is placed into operation, the user must read the instructions. It is important that you, as a Soldier, know how to read these instructions. The most logical place to begin is the marginal information and symbols, where useful information telling about the map is located and explained. All maps are not the same, so it is necessary to examine the marginal information carefully each time a different map is used.

3-1. MARGINAL INFORMATION ON A MILITARY MAP

Figure 3-1 (page 3-4) shows a reduced version of a large-scale topographic map. The circled numbers indicate the items of marginal information that the map user needs to know. These circled numbers correspond to the following listed items.

    a. **Sheet Name (1)**. The sheet name is found in bold print at the center of the top and in the lower left area of the map margin. A map is generally named for the largest settlement contained within the area covered by the sheet, or for the largest natural feature located within the area at the time the map was drawn.

    b. **Sheet Number (2)**. The sheet number is found in bold print in both the upper right and lower left areas of the margin, and in the center box of the adjoining sheets diagram, which is found in the lower right margin. It is used as a reference number to link specific maps to overlays, operations orders, and plans. For maps at 1:100,000 scale and larger, sheet numbers are based on an arbitrary system that makes possible the ready orientation of maps at scales of 1:100,000, 1:50,000, and 1:25,000.

    c. **Series Name (3)**. The map series name is found in bold print in the upper left corner of the margin. The name given to the series is generally that of a major political subdivision such as a state within the United States or a European nation. A map series usually includes a group of similar maps at the same scale and on the same sheet lines or format designed to cover a particular geographic area. It may also be a group of maps that serve a common purpose such as the military city maps.

    d. **Scale (4)**. The scale is found both in the upper left margin after the series name, and in the center of the lower margin. The scale note is a representative fraction that gives the ratio of a map distance to the corresponding distance on the earth's surface. For example, the scale note 1:50,000 indicates that one unit of measure on the map equals 50,000 units of the same measure on the ground.

    e. **Series Number (5)**. The series number is found in both the upper right margin and the lower left margin. It is a sequence reference expressed either as a four-digit numeral (1125) or as a letter, followed by a three- or four-digit numeral (M661, T7110).

    f. **Edition Number (6)**. The edition number is found in bold print in the upper right area of the top margin and the lower left area of the bottom margin. Editions are numbered consecutively; therefore, if you have more than one edition, the highest numbered sheet is the most recent. Most military maps are now published by the NGA, but older editions of maps may have been produced by the U.S. Army Map Service. Still others may have been drawn, at least in part, by the U.S. Army Corps of Engineers, the U.S. Geological Survey, or

FOUO   3-1
18 January 2005

other agencies affiliated or not with the United States or allied governments. The credit line, telling who produced the map, is just above the legend. The map information date is found immediately below the word "LEGEND" in the lower left margin of the map. This date is important when determining how accurately the map data might be expected to match what you will encounter on the ground.

g. **Index to Boundaries (7).** The index to boundaries diagram appears in the lower or right margin of all sheets. This diagram, which is a miniature of the map, shows the boundaries that occur within the map area such as county lines and state boundaries.

h. **Adjoining Sheets Diagram (8).** Maps at all standard scales contain a diagram that illustrates the adjoining sheets. On maps at 1:100,000 and larger scales and at 1:1,000,000 scale, the diagram is called the index to adjoining sheets. It consists of as many rectangles representing adjoining sheets as are necessary to surround the rectangle that represents the sheet under consideration. The diagram usually contains nine rectangles, but the number may vary depending on the locations of the adjoining sheets. All represented sheets are identified by their sheet numbers. Sheets of an adjoining series, whether published or planned, that are at the same scale are represented by dashed lines. The series number of the adjoining series is indicated along the appropriate side of the division line between the series.

i. **Elevation Guide (9).** The elevation guide is normally found in the lower right margin. It is a miniature characterization of the terrain shown. The terrain is represented by bands of elevation, spot elevations, and major drainage features. The elevation guide provides the map reader with a means of quick recognition of major landforms.

j. **Declination Diagram (10).** The declination diagram is located in the lower margin of large-scale maps and indicates the angular relationships of true north, grid north, and magnetic north. On maps at 1:250,000 scale, this information is expressed as a note in the lower margin. In recent edition maps, there is a note indicating the conversion of azimuths from grid to magnetic and from magnetic to grid next to the declination diagram.

k. **Bar Scales (11).** Bar scales are located in the center of the lower margin. They are rulers used to convert map distance to ground distance. Maps have three or more bar scales, each in a different unit of measure. Care should be exercised when using the scales, especially in the selection of the unit of measure that is needed.

l. **Contour Interval Note (12).** The contour interval note is found in the center of the lower margin normally below the bar scales. It states the vertical distance between adjacent contour lines of the map. When supplementary contours are used, the interval is indicated. In recent edition maps, the contour interval is given in meters instead of feet.

m. **Spheroid Note (13).** The spheroid note is located in the center of the lower margin. Spheriods (ellipsoids) have specific parameters that define the X Y Z axis of the earth. The spheriod is an integral part of the datum.

n. **Grid Note (14).** The grid note is located in the center of the lower margin. It gives information pertaining to the grid system used and the interval between grid lines, and it identifies the UTM grid zone number.

o. **Projection Note (15).** The projection system is the framework of the map. For military maps, this framework is of the conformal type; that is, small areas of the surface of the earth retain their true shapes on the projection; measured angles closely approximate true values; and the scale factor is the same in all directions from a point. The projection note is located in the center of the lower margin. (Refer to NGA for the development characteristics of the conformal-type projection systems.)

(1) Between 80 degrees south and 84 degrees north, maps at scales larger than 1:500,000 are based on the transverse Mercator projection. The note reads TRANSVERSE MERCATOR PROJECTION.

(2) Between 80 degrees south and 84 degrees north, maps at 1:1,000,000 scale and smaller are based on standard parallels of the lambert conformal conic projection. The note reads, for example, LAMBERT CONFORMAL CONIC PROJECTIONS 36 DEGREES 40' N AND 39 DEGREES 20' N.

(3) Maps of the polar regions (south of 80 degrees south and north of 84 degrees north) at 1:1,000,000 and larger scales are based on the polar stereographic projection. The note reads POLAR STEREOGRAPHIC PROJECTION.

p. **Vertical Datum Note (16).** The vertical datum note is located in the center of the lower margin. The vertical datum or vertical-control datum is defined as any level surface taken as a surface of reference from which to determine elevations. In the United States, Canada, and Europe, the vertical datum refers to the mean sea level surface. However, in parts of Asia and Africa, the vertical-control datum may vary locally and is based on an assumed elevation that has no connection to any sea level surface. Map readers should habitually check the vertical datum note on maps, particularly if the map is used for low-level aircraft navigation, naval gunfire support, or missile target acquisition.

q. **Horizontal Datum Note (17).** The horizontal datum note is located in the center of the lower margin. The horizontal datum or horizontal-control datum is defined as a geodetic reference point (of which five quantities are known: latitude, longitude, azimuth of a line from this point, and two constants, which are the parameters of reference ellipsoid). These are the basis for horizontal-control surveys. The horizontal-control datum may extend over a continent or be limited to a small local area. Maps and charts produced by NGA are produced on 32 different horizontal-control data. Map readers should habitually check the horizontal datum note on every map or chart, especially adjacent map sheets, to ensure the products are based on the same horizontal datum. If products are based on different horizontal-control data, coordinate transformations to a common datum must be performed. UTM coordinates from the same point computed on different data may differ as much as 900 meters.

r. **Control Note (18).** The control note is located in the center of the lower margin. It indicates the special agencies involved in the control of the technical aspects of all the information that is disseminated on the map.

s. **Preparation Note (19).** The preparation note is located in the center of the lower margin. It indicates the agency responsible for preparing the map.

t. **Printing Note (20).** The printing note is also located in the center of the lower margin. It indicates the agency responsible for printing the map and the date the map was printed. The printing data should not be used to determine when the map information was obtained.

u. **Grid Reference Box (21).** The grid reference box is normally located in the center of the lower margin. It contains instructions for composing a grid reference.

v. **Unit Imprint and Symbol (22).** The unit imprint and symbol is on the left side of the lower margin. It identifies the agency that prepared and printed the map with its respective symbol. This information is important to the map user in evaluating the reliability of the map.

FM 3-25.26

w. **Legend (23).** The legend is located in the lower left margin. It illustrates and identifies the topographic symbols used to depict some of the more prominent features on the map. The symbols are not always the same on every map. Always refer to the legend to avoid errors when reading a map.

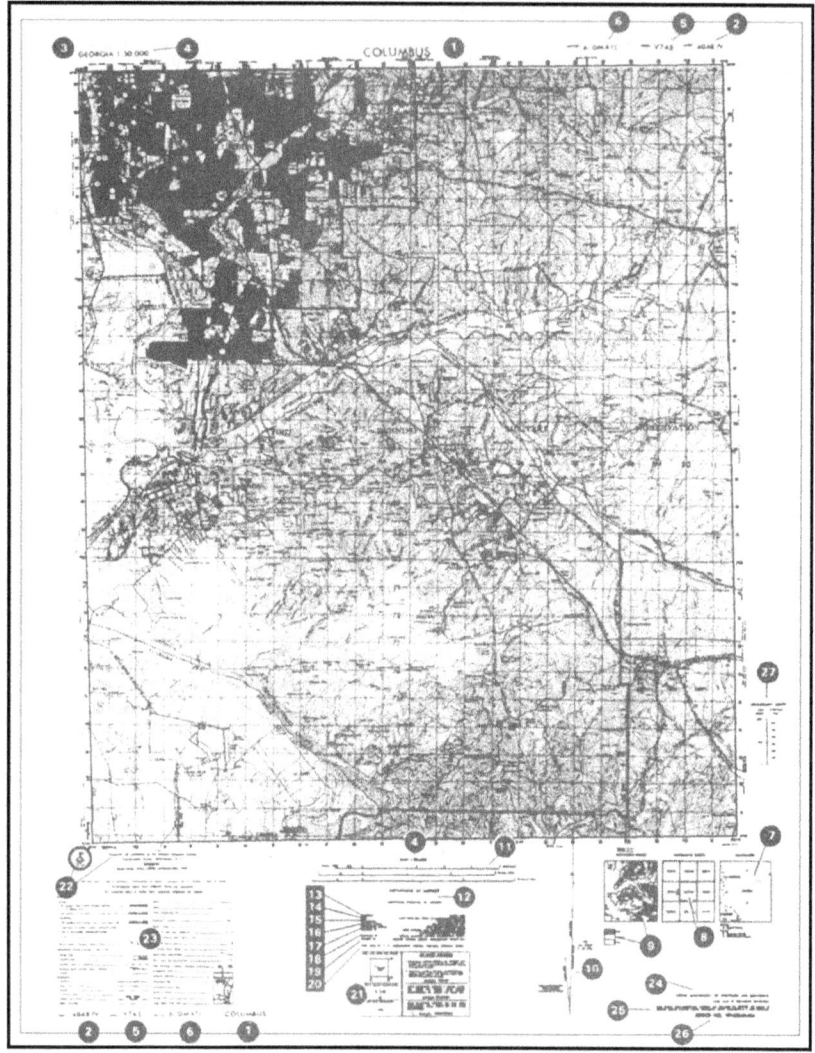

Figure 3-1. Topographical map.

3-2. ADDITIONAL NOTES
Not all maps contain the same items of marginal information. Under certain conditions, special notes and scales may be added to aid the map user.

a. **Glossary.** The glossary is an explanation of technical terms or a translation of terms on maps of foreign areas where the native language is other than English.

b. **Classification.** Certain maps require a note indicating the security classification. This is shown in the upper and lower margins.

c. **Protractor Scale.** The protractor scale may appear in the upper margin on some maps. It is used to lay out the magnetic-grid declination for the map, which, in turn, is used to orient the map sheet with the aid of the lensatic compass.

d. **Coverage Diagram.** On maps at scales of 1:100,000 and larger, a coverage diagram may be used. It is normally in the lower or right margin and indicates the methods by which the map was made, dates of photography, and reliability of the sources. On maps at 1:250,000 scale, the coverage diagram is replaced by a reliability diagram.

e. **Special Notes (24).** A special note is any statement of general information that relates to the mapped area. It is normally found in the lower right margin. For example: This map is red-light readable.

f. **User's Note (25).** The user's note is normally located in the lower right-hand margin. It requests cooperation in correcting errors or omissions on the map. Errors should be marked and the map forwarded to the agency identified in the note.

g. **Stock Number Identification (26).** All maps published by the NGA that are in the Department of the Army map supply system contain stock number identifications that are used in requisitioning map supplies. The identification consists of the words "STOCK NO" followed by a unique designation that is composed of the series number, the sheet number of the individual map and, on recently printed sheets, the edition number. The designation is limited to 15 units (letters and numbers). The first 5 units are allotted to the series number; when the series number is less than 5 units, the letter "X" is substituted as the fifth unit. The sheet number is the next component; however, Roman numerals, which are part of the sheet number, are converted to Arabic numerals in the stock number. The last 2 units are the edition number; the first digit of the edition number is a zero if the number is less than 10. If the current edition number is unknown, the number 01 is used. The latest available edition will be furnished. Asterisks are placed between the sheet number and the edition number when necessary to ensure there are at least 11 units in the stock number.

h. **Conversion Graph (27).** Normally found in the right margin, the conversion graph indicates the conversion of different units of measure used on the map.

3-3. TOPOGRAPHIC MAP SYMBOLS

The purpose of a map is to permit one to visualize an area of the earth's surface with pertinent features properly positioned. The map's legend contains the symbols most commonly used in a particular series or on that specific topographic map sheet. Therefore, the legend should be referred to each time a new map is used. Every effort is made to design standard symbols that resemble the features they represent. If this is not possible, symbols are selected that logically imply the features they portray. For example, an open-pit mining operation is represented by a small black drawing of a crossed hammer and pickax.

a. Ideally, all the features within an area would appear on a map in their true proportion, position, and shape. This, however, is not practical because many of the features would be unimportant and others would be unrecognizable because of their reduction in size.

b. The mapmaker has been forced to use symbols to represent the natural and man-made features of the earth's surface. These symbols resemble, as closely as possible, the actual features as viewed from above. They are positioned in such a manner that the center of the symbol remains in its true location. An exception to this would be the position of a feature adjacent to a major road. If the width of the road has been exaggerated, then the feature is moved from its true position to preserve its relation to the road. (FM 21-31 provides a good description of topographic features and abbreviations that are authorized for use on military maps.)

3-4. MILITARY SYMBOLS

In addition to the topographic symbols used to represent the natural and man-made features of the earth, military personnel require some method for showing identity, size, location, or movement of Soldiers, and military activities and installations. The symbols used to represent these military features are known as military symbols. These symbols are not normally printed on maps because the features and units they represent are constantly moving or changing; military security is also a consideration. They do appear in special maps and overlays (Chapter 7). The map user draws them in, in accordance with proper security precautions. (Refer to FM 101-5-1 for complete information on military symbols.)

3-5. COLORS USED ON A MILITARY MAP

By the fifteenth century, most European maps were carefully colored. Profile drawings of mountains and hills were shown in brown, rivers and lakes in blue, vegetation in green, roads in yellow, and special information in red. A look at the legend of a modern map confirms that the use of colors has not changed much over the past several hundred years. To facilitate the identification of features on a map, the topographical and cultural information is usually printed in different colors. These colors may vary from map to map. On a standard large-scale topographic map, the colors used and the features each represent are—

    a. **Black.** Black indicates cultural (man-made) features such as buildings and roads, surveyed spot elevations, and all labels.

    b. **Red-Brown.** The colors red and brown are combined to identify cultural features, all relief features, nonsurveyed spot elevations, and elevation such as contour lines on red-light readable maps.

    c. **Blue.** Blue identifies hydrography or water features such as lakes, swamps, rivers, and drainage.

    d. **Green.** Green identifies vegetation with military significance such as woods, orchards, and vineyards.

    e. **Brown.** Brown identifies all relief features and elevation such as contours on older edition maps and cultivated land on red-light readable maps.

    f. **Red.** Red classifies cultural features, such as populated areas, main roads, and boundaries, on older maps.

    g. **Other.** Occasionally, other colors may be used to show special information. These are indicated in the marginal information as a rule.

CHAPTER 4
GRIDS

This chapter covers how to determine and report positions on the ground in terms of their locations on a map. Knowing where you are (position fixing) and being able to communicate that knowledge is crucial to successful land navigation as well as to the effective employment of direct and indirect fire, tactical air support, and medical evacuation. It is essential for valid target acquisition; accurate reporting of NBC contamination and various danger areas; and obtaining emergency resupply. Few factors contribute as much to the survivability of troops and equipment and to the successful accomplishment of a mission as always knowing where you are. The chapter includes explanations of geographical coordinates, Universal Transverse Mercator grids, the military grid reference system, and the use of grid coordinates.

4-1. REFERENCE SYSTEM

In a city, it is quite simple to find a location; the streets are named and the buildings have numbers. The only thing needed is the address. However, finding locations in undeveloped areas or in unfamiliar parts of the world can be a problem. To cope with this problem, a uniform and precise system of referencing has been developed.

4-2. GEOGRAPHIC COORDINATES

One of the oldest systematic methods of location is based upon the geographic coordinate system. By drawing a set of east-west rings around the globe (parallel to the equator), and a set of north-south rings crossing the equator at right angles and converging at the poles, a network of reference lines is formed from which any point on the earth's surface can be located.

    a. The distance of a point north or south of the equator is known as its latitude. The rings around the earth parallel to the equator are called parallels of latitude or simply parallels. Lines of latitude run east-west but north-south distances are measured between them.

    b. A second set of rings around the globe at right angles to lines of latitude and passing through the poles are known as meridians of longitude or simply meridians. One meridian is designated as the prime meridian. The prime meridian of the system we use runs through Greenwich, England and is known as the Greenwich meridian. The distance east or west of a prime meridian to a point is known as its longitude. Lines of longitude (meridians) run north-south but east-west distances are measured between them (Figures 4-1 and 4-2, page 4-2).

FM 3-25.26

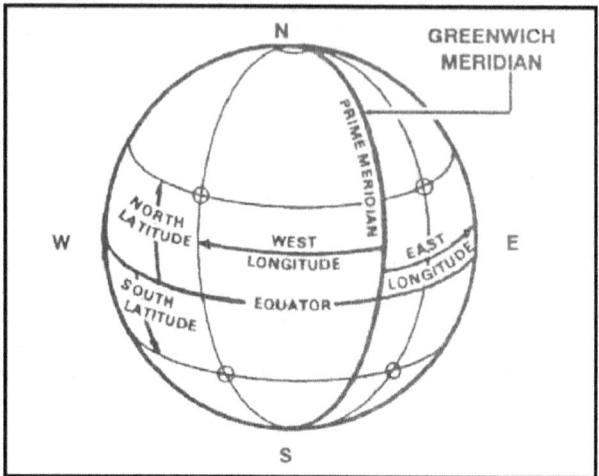

Figure 4-1. Prime meridian and equator.

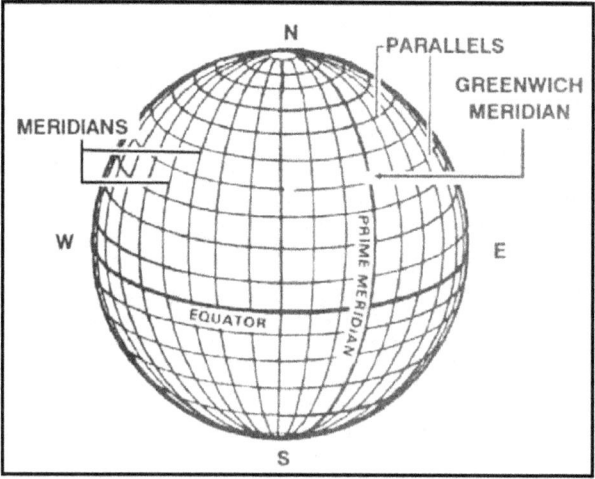

Figure 4-2. Reference lines.

c. Geographic coordinates are expressed in angular measurement. Each circle is divided into 360 degrees, each degree into 60 minutes, and each minute into 60 seconds. The degree is symbolized by °, the minute by ', and the second by ".

(1) Starting with 0° at the equator, the parallels of latitude are numbered to 90° both north and south. The extremities are the north pole at 90° north latitude and the south pole at 90° south latitude. Latitude can have the same numerical value north or south of the equator, so the direction N or S must always be given.

4-2   FOUO
         18 January 2005

(2) Starting with 0° at the prime meridian, longitude is measured both east and west around the world. Lines east of the prime meridian are numbered to 180° and identified as east longitude; lines west of the prime meridian are numbered to 180° and identified as west longitude. The direction E or W must always be given. The line directly opposite the prime meridian, 180°, may be referred to as either east or west longitude.

(3) The values of geographic coordinates, being in units of angular measure, will mean more if they are compared with more familiar units of measure. At any point on the earth, the ground distance covered by one degree of latitude is about 111 kilometers (69 miles); one second is equal to about 30 meters (100 feet). The ground distance covered by one degree of longitude at the equator is also about 111 kilometers, but decreases as one moves north or south, until it becomes zero at the poles. For example, one second of longitude represents about 30 meters (100 feet) at the equator; but at the latitude of Washington, DC, one second of longitude is about 24 meters (78 feet). Latitude and longitude are illustrated in Figure 4-3.

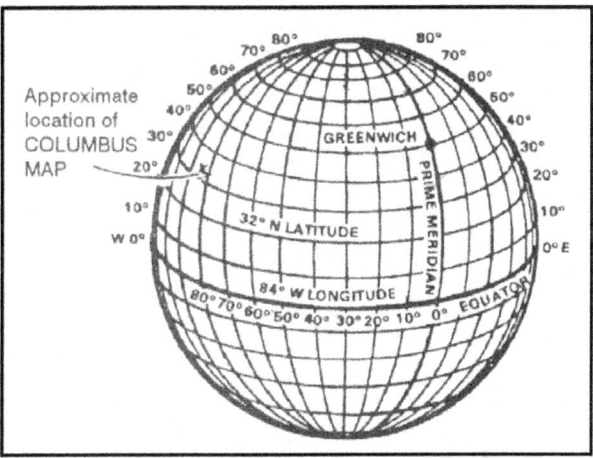

Figure 4-3. Latitude and longitude.

d. Geographic coordinates appear on all standard military maps; on some they may be the only method of locating and referencing the location of a point. The four lines that enclose the body of the map (neatlines) are latitude and longitude lines. Their values are given in degrees and minutes at each of the four corners.

(1) On a portion of the Columbus map (Figure 4-4, page 4-5), the figures 32°15' and 84°45' appear at the lower right corner. The bottom line of this map is latitude 32°15'00"N, and the line running up the right side is longitude 84°45'00"W.

(2) In addition to the latitude and longitude given for the four corners, there are small tick marks at regularly spaced intervals along the sides of the map, extending into the body of the map. Each of these tick marks is identified by its latitude or longitude value.

(3) Near the top of the right side of the map is a tick mark with the number 20'. The full value for this tick marks is 32°20'00" of latitude. At one-third and two-thirds of the distance across the map from the 20' tick mark will be found a cross tick mark (grid squares GL 0379

and FL 9679) and at the far side another 20' tick mark. By connecting the tick marks and crosses with straight lines, a 32°20'00" line of latitude can be added to the map. This procedure is also used to locate the 32°25'00" line of latitude. For lines of longitude, the same procedure is followed using the tick marks along the top and bottom edges of the map.

e. After the parallels and meridians have been drawn, the geographic interval (angular distance between two adjacent lines) must be determined. Examination of the values given at the tick marks gives the interval. For most maps of scale 1:25,000, the interval is 2'30". For the Columbus map and most maps of scale 1:50,000, it is 5'00". The geographic coordinates of a point are found by dividing the sides of the geographic square in which the point is located into the required number of equal parts. If the geographic interval is 5'00" and the location of a point is required to the nearest second, each side of the geographic square must be divided into 300 equal parts (5'00" = 300"), each of which would have a value of one second. Any scale or ruler that has 300 equal divisions and is as long as or longer than the spacing between the lines may be used.

f. The following steps will determine the geographic coordinates of Wilkinson Cemetery (northwest of the town of Cusseta) on the Columbus map.

(1) Draw on the map the parallels and meridians that enclose the area around the cemetery.

(2) Determine the values of the parallels and meridians where the point falls.

Latitude 32°15'00" and 32°20'00".

Longitude 84°45'00" and 84°50'00".

(3) Determine the geographic interval (5'00" = 300").

(4) Select a scale that has 300 small divisions or multiples thereof (300 divisions, one second each; 150 divisions, two seconds each; 75 divisions, four seconds each, and so forth).

(5) To determine the latitude—

(a) Place the 0 of the scale on the latitude of the lowest number value (32°15'00") and the 300 of the scale on the highest numbered line (32°20'00") (1, Figure 4-4).

(b) Keeping the 0 and 300 on the two lines, slide the scale (2, Figure 4-4) along the parallels until the Wilkinson Cemetery symbol is along the edge of the numbered scale.

(c) Read the number of seconds from the scale (3, Figure 4-4), about 246.

(d) Convert the number of seconds to minutes and seconds (246" = 4'06") and add to the value of the lower numbered line (32°15'00" + 4'06" = 32°19'06") (4, Figure 4-4).

(e) The latitude is 32°19'06", but this information alone is not enough. The latitude 32°19'06" could be either north or south of the equator, so the letter N or S must be added to the latitude. To determine whether it is N or S, look at the latitude values at the edge of the map and find the direction in which they become larger. If they are larger going north, use N; if they are larger going south, use S. The latitude for the cemetery is 32°19'06"N.

(6) To determine the longitude, repeat the same steps but measure between lines of longitude and use E and W. The geographic coordinates of Wilkinson Cemetery should be about 32°19'06"N and 84°47'32"W (Figure 4-5, page 4-6).

g. Many of the same steps are followed to locate a point on the Columbus map when knowing the geographic coordinates (Figure 4-6, page 4-7). To locate 32°25'28"N and 84°50'56"W, first find the geographic lines within which the point falls: latitude 32°25'00" and 32°30'0"; and longitude 84°50'00" and 84°55'00". Subtract the lower latitude or longitude from the higher latitude or longitude.

(1) Place the 0 of the scale on the 32°25'00" line and the 300 on the 32°30'00". Make a mark at the number 28 on the scale (the difference between the lower and higher latitude).

(2) Place the 0 of the scale on the 84°50'00" line and the 300 on the 84°50'55". Make a mark at the number 56 on the scale (the difference between the lower and higher longitude).

(3) Draw a vertical line from the mark at 56 and a horizontal line from the mark at 28; they intersect at 32 25'28"N and 84 50'56"W.

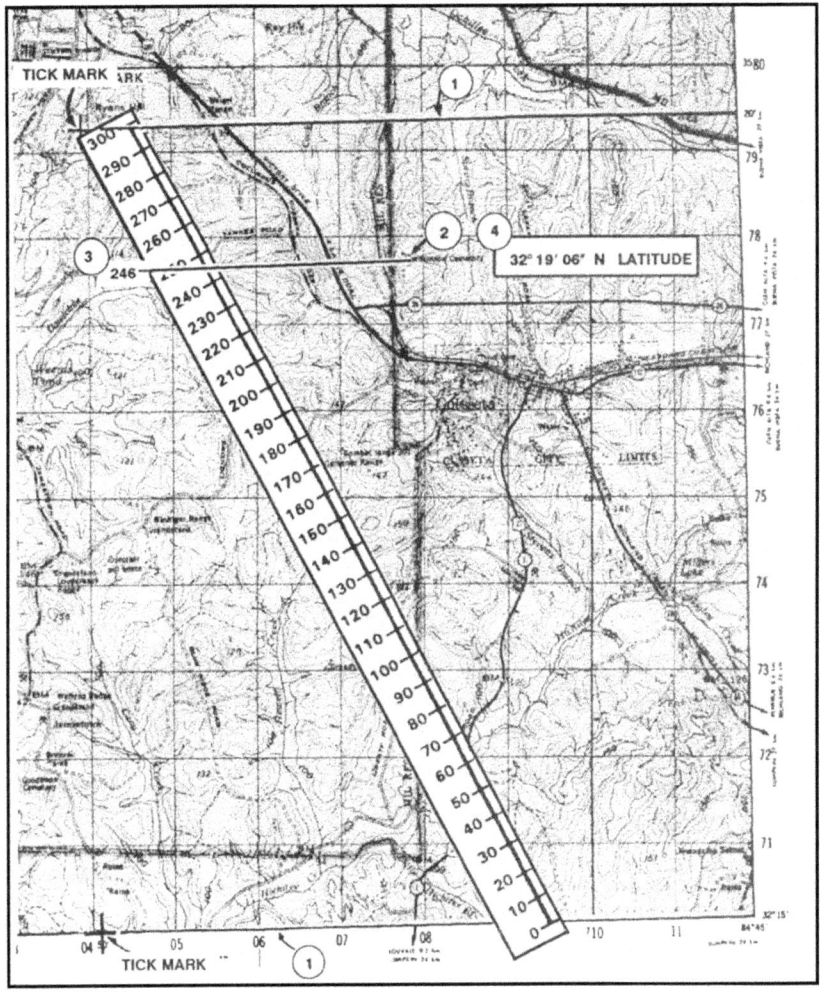

Figure 4-4. Determining latitude.

FM 3-25.26

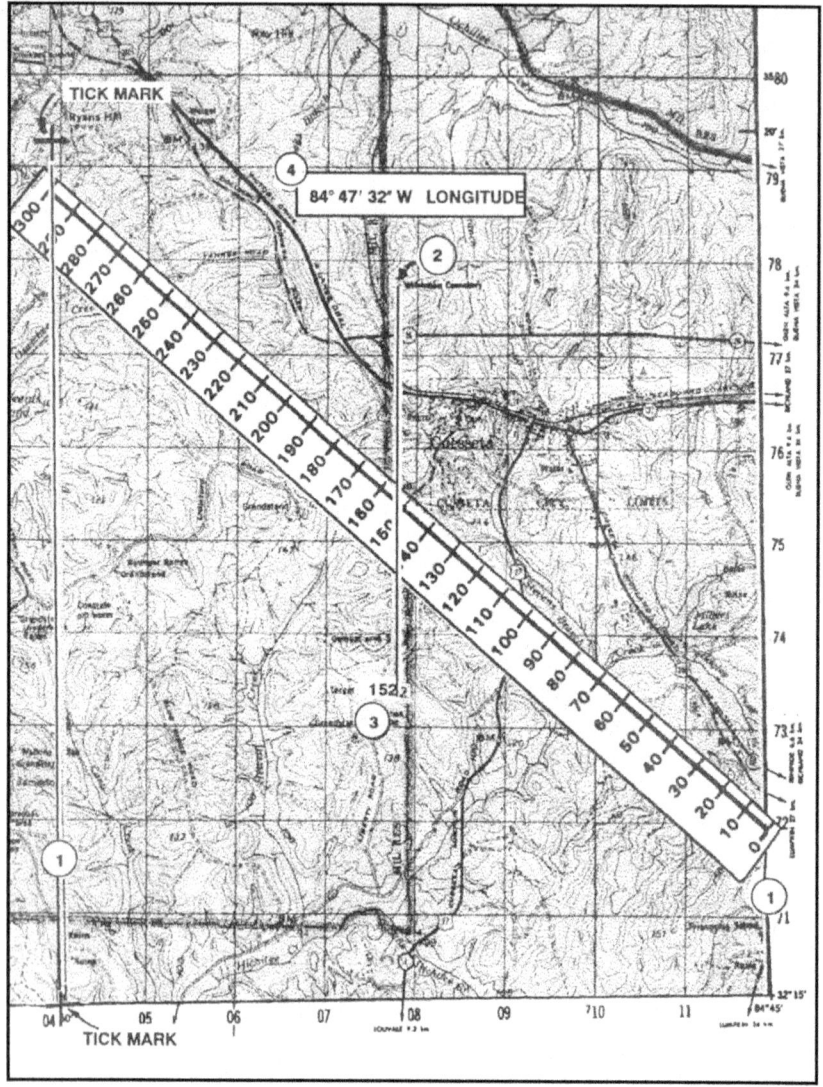

Figure 4-5. Determining longitude.

C1, FM 3-25.26

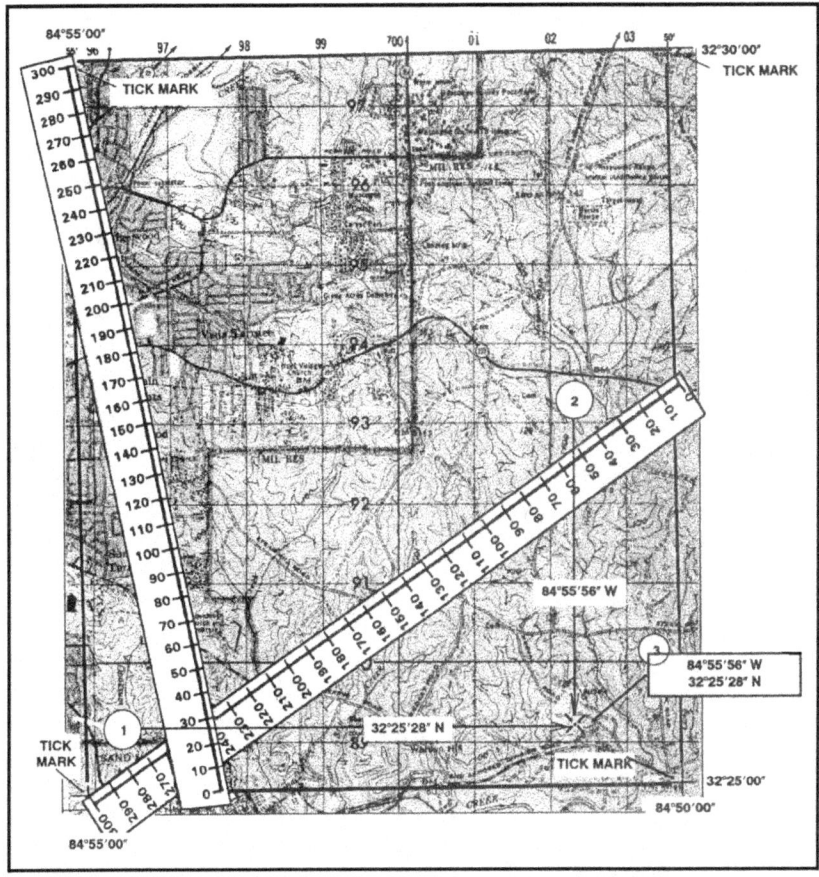

Figure 4-6. Determining geographic coordinates.

h. If you do not have a scale or ruler with 300 equal divisions or a map whose interval is other than 5'00", use the proportional parts method. Following the steps determines the geographic coordinates of horizontal control station 141.

(1) Locate horizontal control station 141 in grid square (GL0784) (Figure 4-7, page 4-8).

(2) Find a cross in grid square GL0388 and a tick mark in grid square GL1188 with 25'.

(3) Find another cross in grid square GL0379 and a tick mark in grid square GL1179 with 20'.

(4) Enclose the control station by connecting the crosses and tick marks. The control station is between 20' and 25'.

(5) With a boxwood scale, measure the distance from the bottom line to the top line that encloses the area around the control station on the map (total distance).

30 August 2006 4-7

FOUO

C1, FM 3-25.26

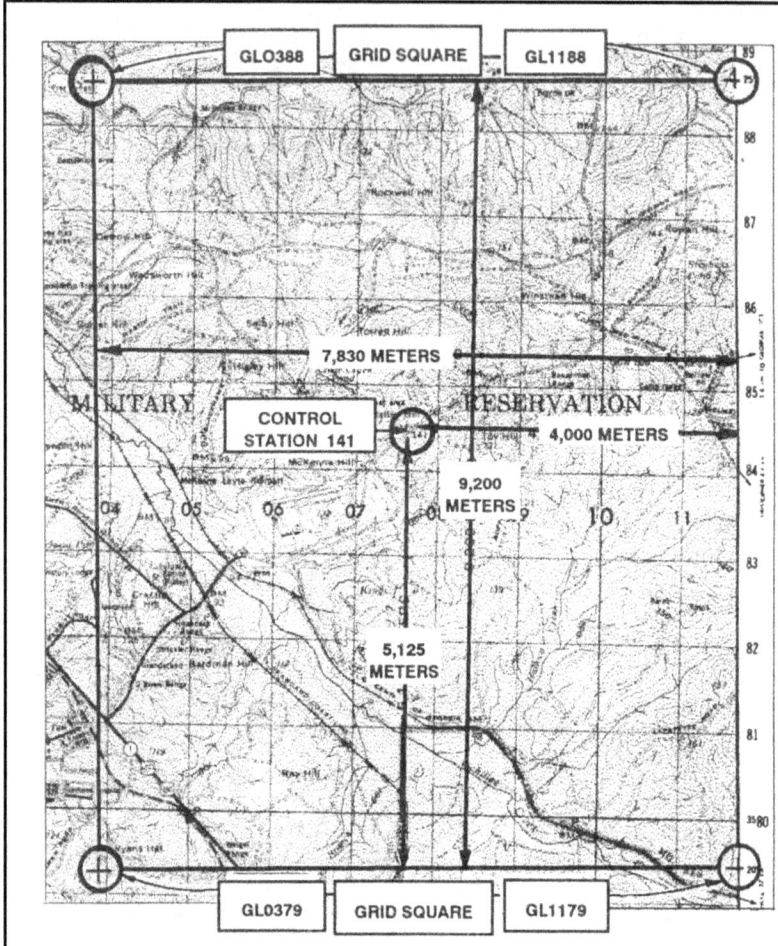

Figure 4-7. Using the proportional parts method.

(6) Measure the partial distance from the bottom line to the center of the station. These straight-line distances are in direct proportion to the minutes and s of latitude and are used to set up a ratio.

(7) The total distance is 9,200 meters, and the partial distance is 5,125 meters.

(8) With the two distances and the five-minute interval converted to seconds determine the minutes and seconds of latitude using the following formula:
1. 5,125 x 300 = 1,537,500
2. 1,537,500 ÷ 9,200 = 167
3. 167 ÷ 60 = 2'47"
*4. Add 2'47" to 32°20'00" = 32°22'47"

(9) Follow the same procedures to determine minutes and seconds of longitude.
(10) The total distance is 7,830 meters, and the partial distance is 4,000 meters.
1. 4,000 x 300 = 1,200,000
2. 1,200,000 ÷ 7,830 = 153
3. 153 ÷ 60 = 2'33"
4. Add 2'33" to 84°45' = 84°47'33"N

(11) The geographic coordinates of horizontal control station 141 in grid square GL0784 are 32°22'47"N latitude and 84°47'33"W longitude.

NOTE: When computing formulas, you must round off totals to the nearest whole number in step 2. In step 3, convert the fraction to seconds by multiplying the fraction by 60 and rounding off if the total is not a whole number.

i. The maps made by some nations do not have their longitude values based on the prime meridian that passes through Greenwich, England. Table 4-1 shows the prime meridians that may be used by other nations. When these maps are issued to our soldiers, a note usually appears in the marginal information giving the difference between our prime meridian and the one used on the map.

CITY, COUNTRY	PRIME MERIDIAN
Amsterdam, Netherlands	4°53'01"E
Athens, Greece	23°42'59"E
Batavia (Djakarta), Indonesia	106°48'28"E
Bern, Switzerland	7°26'22"E
Brussels, Belgium	4°22'06"E
Copenhagen, Denmark	12°34'40"E
Ferro (Hierro), Canary Islands	17°39'46"W
Helsinki, Finland	24°53'17"E
Istanbul, Turkey	28°58'50"E
Lisbon, Portugal	9°07'55"W
Madrid, Spain	3°41'15"W
Oslo, Norway	10°43'23"E
Paris, France	2°20'14"E
Pulkovo, Russia	30°19'39"E
Rome, Italy	12°27'08"E
Stockholm, Sweden	18°03'30"E
Tirane, Albania	19°46'45"E

Table 4-1. Table of prime meridians.

4-3. MILITARY GRIDS

An examination of the transverse Mercator projection, which is used for large-scale military maps, shows that most lines of latitude and longitude are curved lines. The quadrangles formed by the intersection of these curved parallels and meridians are of different sizes and shapes, complicating the location of points and the measurement of directions. To aid these essential operations, a rectangular grid is superimposed upon the projection. This grid (a

FM 3-25.26

series of straight lines intersecting at right angles) furnishes the map reader with a system of squares similar to the block system of most city streets. The dimensions and orientation of different types of grids vary, but three properties are common to all military grid systems: one, they are true rectangular grids; two, they are superimposed on the geographic projection; and three, they permit linear and angular measurements.

a. **Universal Transverse Mercator (UTM) Grid**. The UTM grid system was adopted by the U.S. Army in 1947 for designating rectangular coordinates on large-scale military maps. The UTM is currently used by the United States and NATO armed forces. With the advent of inexpensive GPS receivers, many other map users are adopting the UTM grid system for coordinates that are simpler to use than latitude and longitude. The UTM grid was designed to cover that part of the world between latitude 84°N and latitude 80°S, and, as its name implies, is imposed on the transverse Mercator projection. Each of the 60 zones (6 degrees wide) into which the globe is divided for the grid has its own origin at the intersection of its central meridian and the equator (Figure 4-8). The grid is identical in all 60 zones. Base values (in meters) are assigned to the central meridian and the equator, and the grid lines are drawn at regular intervals parallel to these two base lines. With each grid line assigned a value denoting its distance from the origin, the problem of locating any point becomes progressively easier. Normally, it would seem logical to assign a value of zero to the two base lines and measure outward from them. This, however, would require either that directions—N, S, E, or W—be always given with distances, or that all points south of the equator or west of the central meridian have negative values. This inconvenience is eliminated by assigning "false values" to the base lines, resulting in positive values for all points within each zone. Distances are always measured RIGHT and UP (east and north as the reader faces the map), and the assigned values are called "false easting" and "false northing." (Figure 4-9). The false easting value for each central meridian is 500,000 meters, and the false northing value for the equator is 0 meters when measuring in the northern hemisphere and 10,000,000 meters when measuring in the southern hemisphere. (The use of the UTM grid for point designation will be discussed in detail in paragraph 4-4.)

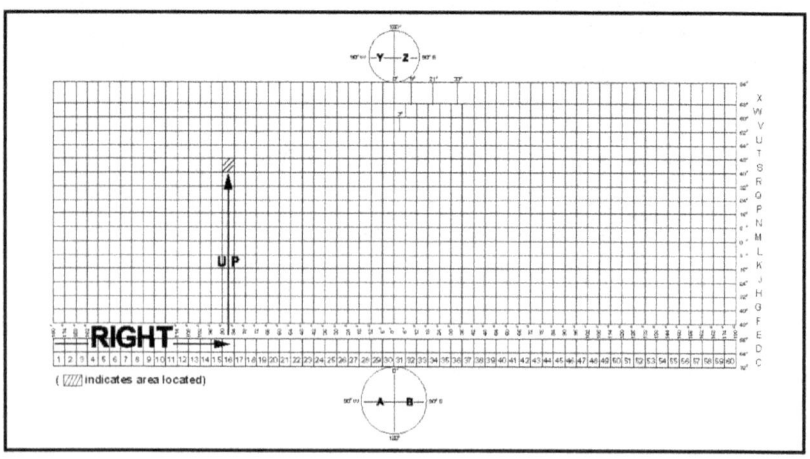

Figure 4-8. UTM grid zone location.

FOUO

18 January 2005

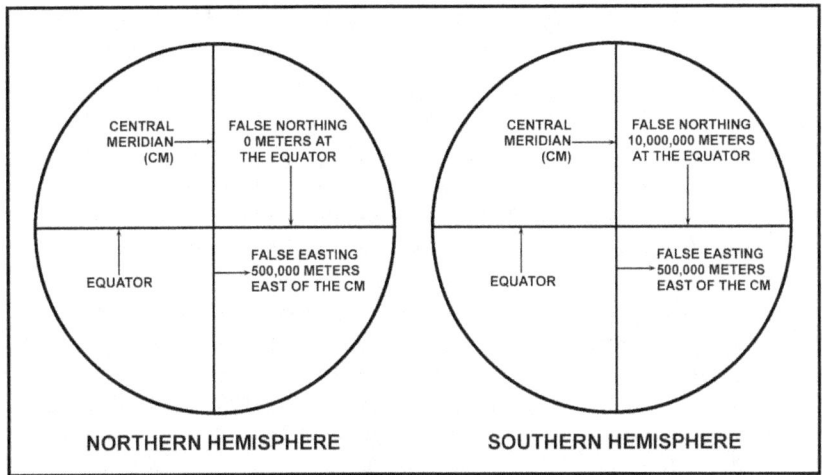

Figure 4-9. False eastings and northings for the UTM grid.

b. **Universal Polar Stereographic (UPS) Grid.** The UPS grid is used to represent the polar regions (Figure 4-10, page 4-12).

(1) *North Polar Area.* The origin of the UPS grid applied to the north polar area is the north pole. The "north-south" base line is the line formed by the 0-degree and 180-degree meridians; the "east-west" base line is formed by the two 90-degree meridians.

(2) *South Polar Area.* The origin of the UPS grid in the south polar area is the south pole. The base lines are similar to those of the north polar area.

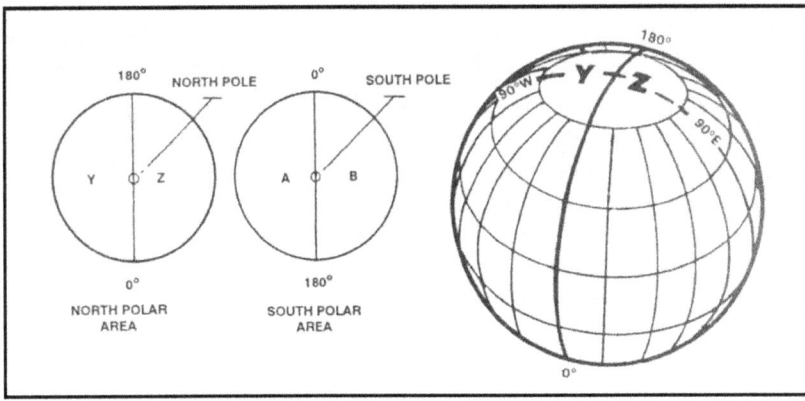

Figure 4-10. Grid zone designation for UPS grid.

4-4. UNITED STATES ARMY MILITARY GRID REFERENCE SYSTEM

This grid reference system is designated for use with the UTM and UPS grids. The coordinate value of points in these grids could contain as many as 15 digits if numerals alone were used. The U.S. military grid reference system reduces the length of written coordinates by substituting single letters for several numbers. Using the UTM and the UPS grids, it is possible for the location of a point (identified by numbers alone) to be in many different places on the surface of the earth. With the use of the military grid reference system, there is no possibility of this happening.

   a. **Grid Zone Designation.** The world is divided into 60 grid zones, which are large, regularly shaped geographic areas, each of which is given a unique identification called the grid zone designation.

   (1) *UTM Grid.* The first major breakdown is the division of each zone into areas 6° wide by 8° high and 6° wide by 12° high. Remember, for the transverse Mercator projection, the earth's surface between 80°S and 84°N is divided into 60 north-south zones, each 6° wide. These zones are numbered from west to east, 1 through 60, starting at the 180° meridian. This surface is divided into 20 east-west rows in which 19 are 8° high and 1 row at the extreme north is 12° high. These rows are then lettered, from south to north, C through X (I and O were omitted). Any 6° by 8° zone or 6° by 12° zone is identified by giving the number and letter of the grid zone and row in which it lies. These are read RIGHT and UP so the number is always written before the letter. This combination of zone number and row letter constitutes the grid zone designation. Columbus lies in zone 16 and row S, or in grid zone designation 16S (Figure 4-8).

   (2) *UPS Grid.* The remaining letters of the alphabet—A, B, Y, and Z—are used for the UPS grids. Each polar area is divided into two zones separated by the 0-180° meridian. In the south polar area, the letter A is the grid zone designation for the area west of the 0-180° meridian, and B for the area to the east. In the north polar area, Y is the grid zone designation for the western area and Z for the eastern area (Figure 4-10).

   b. **100,000-Meter Square.** Between 84°N and 80°S, each 6° by 8° or 6° by 12° zone is covered by 100,000-meter squares that are identified by the combination of two alphabetical letters. This identification is unique within the area covered by the grid zone designation. The first letter is the column designation; the second letter is the row designation (Figure 4-11). The north and south polar areas are also divided into 100,000-meter squares by columns and rows. The 100,000-meter square identification letters are located in the grid reference box in the lower margin of the map.

PLATE 12

	96°						90°			500,000m			84°	
	QV	TQ	UQ	VQ	WQ	XQ	YQ	BV	CV	DV	EV	FV	GV	KQ
QU	TP	UP	VP	WP	XP	YP	BU	CU	DU	EU	FU	GU	KP
QT	TN	UN	VN	WN	XN	YN	BT	CT	DT	ET	FT	GT	KN
QS	TM	UM	VM	WM	XM	YM	BS	CS	DS	ES	FS	GS	KM
QR	TL	UL	VL	WL	XL	YL	BR	CR	DR	ER	FR	GR	K
QQ	TK	UK	VK	WK	XK	YK	BQ	CQ	DQ	EQ	FQ	GQ	K
QP	TJ	UJ	VJ	WJ	XJ	YJ	BP	CP	DP	EP	FP	GP	K
QN	TH	UH	VH	WH	XH	YH	BN	CN	DN	EN	FN	GN	K
QM	TG	UG	VG	WG	XG	YG	BM	CM	DM	EM	FM	GM	K
QL	TF	UF	VF	WF	XF	YF	BL	CL	DL	EL	FL	GL	K

Figure 4-11. Grid zone designation and 100,000-meter square identification.

c. **Grid Coordinates**. We have now divided the earth's surface into 6° by 8° quadrangles, and covered these with 100,000-meter squares. The military grid reference of a point consists of the numbers and letters indicating in which of these areas the point lies, plus the coordinates locating the point to the desired position within the 100,000-meter square. The next step is to tie in the coordinates of the point with the larger areas. To do this, you must understand the following.

(1) **Grid Lines**. The regularly spaced lines that make the UTM and the UPS grid on any large-scale maps are divisions of the 100,000-meter square; the lines are spaced at 10,000- or 1,000-meter intervals (Figure 4-12, page 4-14). Each of these lines is labeled at both ends of the map with its false easting or false northing value, showing its relation to the origin of the zone. Two digits of the values are printed in large type, and these same two digits appear at intervals along the grid lines on the face of the map. These are called the principal digits, and represent the 10,000 and 1,000 digits of the grid value. They are of major importance to the map reader because they are the numbers he will use most often for referencing points. The smaller digits complete the UTM grid designation.

FM 3-25.26

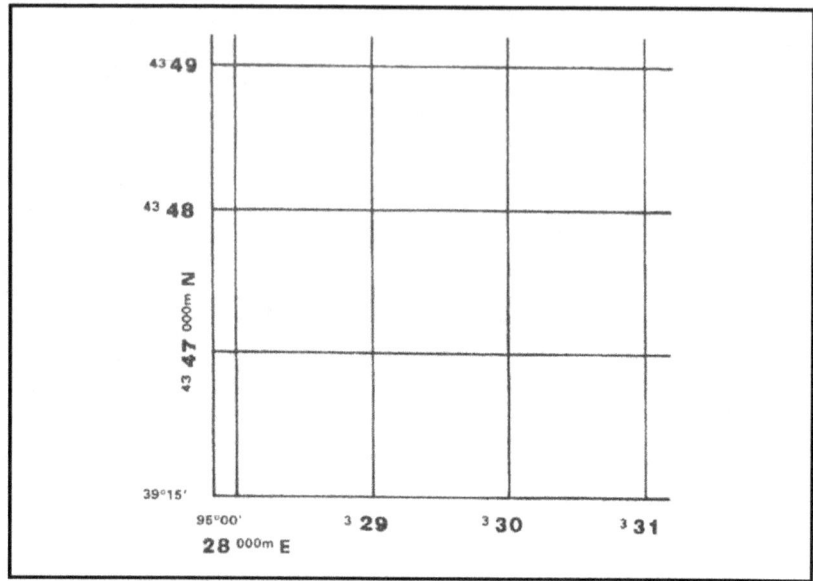

Figure 4-12. Grid lines.

EXAMPLE: The first grid line north of the south-west corner of the Columbus map is labeled 3570000m N. This means its false northing (distance north of the equator) is 3,570,000 meters. The principal digits, 70, identify the line for referencing points in the northerly direction. The smaller digits, 35, are part of the false coordinates and are rarely used. The last three digits, 000, of the value are omitted. Therefore, the first grid line east of the south-west corner is labeled 689000m E. The principal digits, 89, identify the line for referencing points in the easterly direction (Figure 4-13).

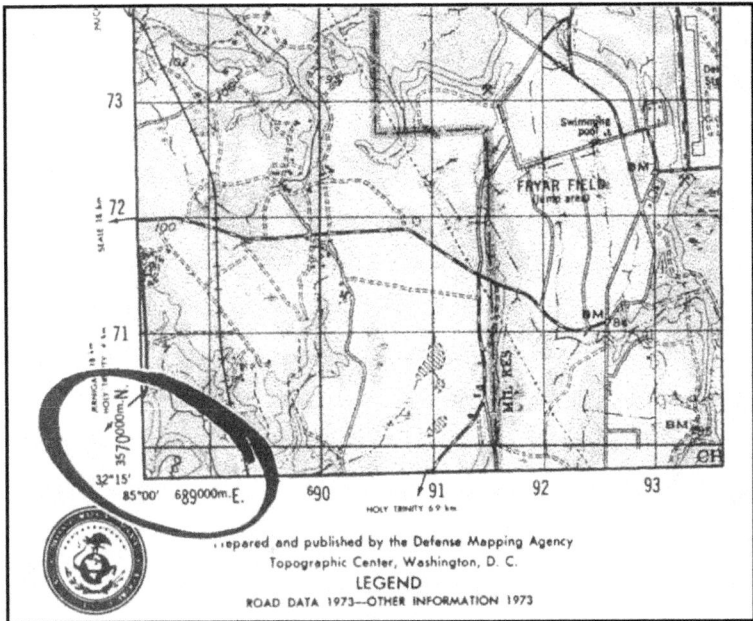

Figure 4-13. Columbus map, southwest corner.

(2) *Grid Squares.* The north-south and east-west grid lines intersect at 90°, forming grid squares. Normally, the size of one of these grid squares on large-scale maps is 1,000 meters (1 kilometer).

(3) *Grid Coordinate Scales.* The primary tool for plotting grid coordinates is the grid coordinate scale. The grid coordinate scale divides the grid square more accurately than can be done by estimation, and the results are more consistent. When used correctly, it presents less chance for making errors. GTA 5-2-12 contains four types of coordinate scales (Figure 4-14, page 4-16).

FM 3-25.26

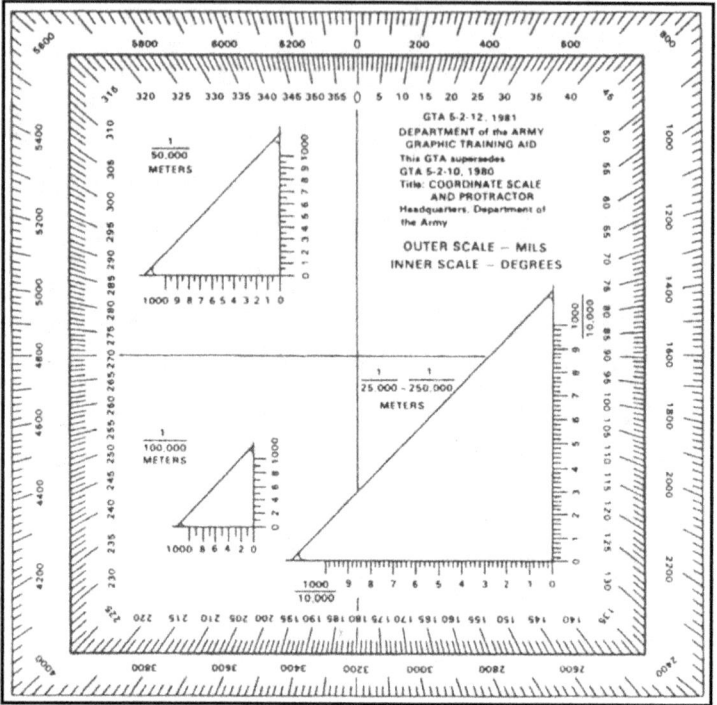

Figure 4-14. Coordinate scales.

(a) The 1:25,000/1:250,000 (lower right in figure) can be used in two different scale maps, 1:25,000 or 1:250,000. The 1:25,000 scale subdivides the 1,000-meter grid block into 10 major subdivisions, each equal to 100 meters. Each 100-meter block has five graduations, each equal to 20 meters. Points falling between the two graduations can be read accurately by the use of estimation. These values are the fourth and eighth digits of the coordinates. Likewise, the 1:250,000 scale is subdivided into 10 major subdivisions, each equal to 1,000 meters. Each 1,000-meter block has five graduations, each equal to 200 meters. Points falling between two graduations can be read approximately by the use of estimation.

(b) The 1:50,000 scale (upper left in Figure 4-14) subdivides the 1,000-meter block into 10 major subdivisions, each equal to 100 meters. Each 100-meter block is then divided in half. Points falling between the graduations must be estimated to the nearest 10 meters for the fourth and eighth digits of the coordinates.

(c) The 1:100,000 scale (lower left in Figure 4-14) subdivides the 1,000-meter grid block into five major subdivisions of 200 meters each. Each 200-meter block is then divided in half at 100-meter intervals.

4-5. LOCATING A POINT USING GRID COORDINATES

Based on the military principle for reading maps (RIGHT and UP), locations on the map can be determined by grid coordinates. The number of digits represents the degree of precision to which a point has been located and measured on a map—the more digits the more precise the measurement.

a. **Without a Coordinate Scale.** In order to determine grids without a coordinate scale, the reader simply refers to the north-south grid lines numbered at the bottom margin of any map. Then he reads RIGHT to the north-south grid line that precedes the desired point (this first set of two digits is the RIGHT reading). Then by referring to the east-west grid lines numbered at either side of the map, the map reader moves UP to the east-west grid line that precedes the desired point (these two digits are the UP reading). Coordinates 1484 locate the 1,000-meter grid square in which point X is located; the next square to the right would be 1584; the next square up would be 1485, and so forth (Figure 4-15). To locate the point to the nearest 100 meters, use estimation. By mentally dividing the grid square in tenths, estimate the distance from the grid line to the point in the same order (RIGHT and UP). Give complete coordinate RIGHT, then complete coordinate UP. Point X is about two-tenths or 200 meters to the RIGHT into the grid square and about seven-tenths or 700 meters UP. The coordinates to the nearest 100 meters are 142847.

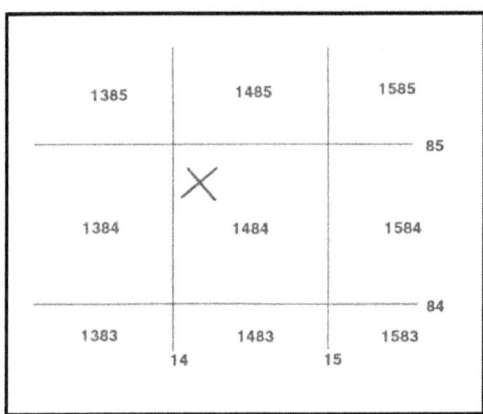

Figure 4-15. Determining grids without coordinate point.

b. **With a Coordinate Scale.** In order to use the coordinate scale for determining grid coordinates, the map user has to make sure that the appropriate scale is being used on the corresponding map, and that the scale is right side up. To ensure the scale is correctly aligned, place it with the zero-zero point at the lower left corner of the grid square. Keeping the horizontal line of the scale directly on top of the east-west grid line, slide it to the right until the vertical line of the scale touches the point for which the coordinates are desired (Figure 4-16, page 4-18). When reading coordinates, examine the two sides of the coordinate scale to ensure that the horizontal line of the scale is aligned with the east-west grid line, and the vertical line of the scale is parallel with the north-south grid line. The scale is used when precision of more than 100 meters is required. To locate the point to the nearest 10 meters, measure the hundredths of a grid square RIGHT and UP from the grid lines to the point. Point X is about 17 hundredths or 170 meters RIGHT and 84 hundredths or 840 meters UP. The coordinates to the nearest 10 meters are 14178484.

c. **Recording and Reporting Grid Coordinates**. Coordinates are written as one continuous number without spaces, parentheses, dashes, or decimal points; they must always contain an even number of digits. Therefore, whoever is to use the written coordinates must know where to make the split between the RIGHT and UP readings. It is a military requirement that the 100,000-meter square identification letters be included in any point designation. Normally, grid coordinates are determined to the nearest 100 meters (six digits) for reporting locations. With practice, this can be done without using plotting scales. The location of targets and other point locations for fire support are determined to the nearest 10 meters (eight digits).

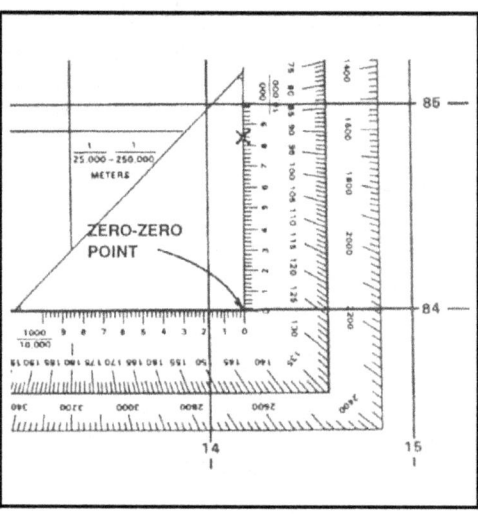

Figure 4-16. Placing a coordinate scale on a grid.

NOTE: Care should be exercised by the map reader using the coordinate scale when the desired point is located within the zero-zero point and the number 1 on the scale. Always prefix a zero if the hundredths reading is less than 10. In Figure 4-17, the desired point should be reported as 14818407.

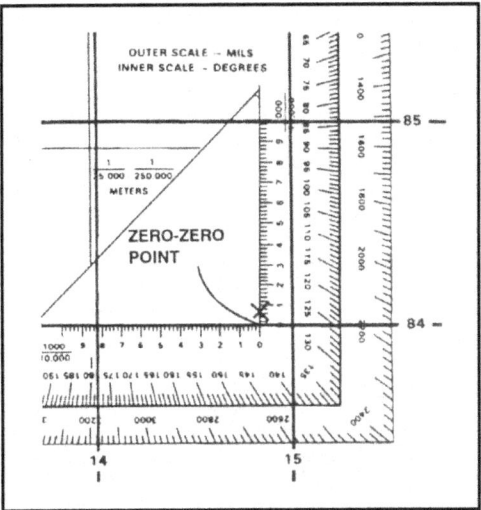

Figure 4-17. Zero-zero point.

NOTE: Special care should be exercised when recording and reporting coordinates. Transposing numbers or making errors could be detrimental to military operations.

4-6. LOCATING A POINT USING THE U.S. ARMY MILITARY GRID REFERENCE SYSTEM

There is only one rule to remember when reading or reporting grid coordinates—always read to the RIGHT and then UP. The first half of the reported set of coordinate digits represents the left-to-right (easting) grid label, and the second half represents the label as read from the bottom to top (northing). The grid coordinates may represent the location to the nearest 10-, 100-, or 1,000-meter increment.

* a. **Grid Zone.** The number 16 locates a point within zone 16, which is an area 6° wide and extends between 80°S latitude and 84°N latitude (Figure 4-8, page 4-10).

* b. **Grid Zone Designation.** The number and letter combination, 16S, further locates a point within the grid zone designation 16S, which is a quadrangle 6° wide by 8° high. There are 19 of these quads in zone 16. Quad X, which is located between 72°N and 84°N latitude, is 12° high (Figure 4-8, page 4-10).

* c. **100,000-Meter Square Identification.** The addition of two more letters locates a point within the 100,000-meter grid square. Thus 16SGL (Figure 4-11, page 4-13) locates the point within the 100,000-meter square GL in the grid zone designation 16S. (For information on the lettering system of 100,000-meter squares, see TM 5-241-1.)

d. **10,000-Meter Square.** The breakdown of the U.S. Army military grid reference system continues as each side of the 100,000-meter square is divided into 10 equal parts. This division produces lines that are 10,000 meters apart. Thus the coordinates 16SGL08 would locate a point as shown in Figure 4-18. The 10,000-meter grid lines appear as index (heavier) grid lines on maps at 1:100,000 and larger.

C1, FM 3-25.26

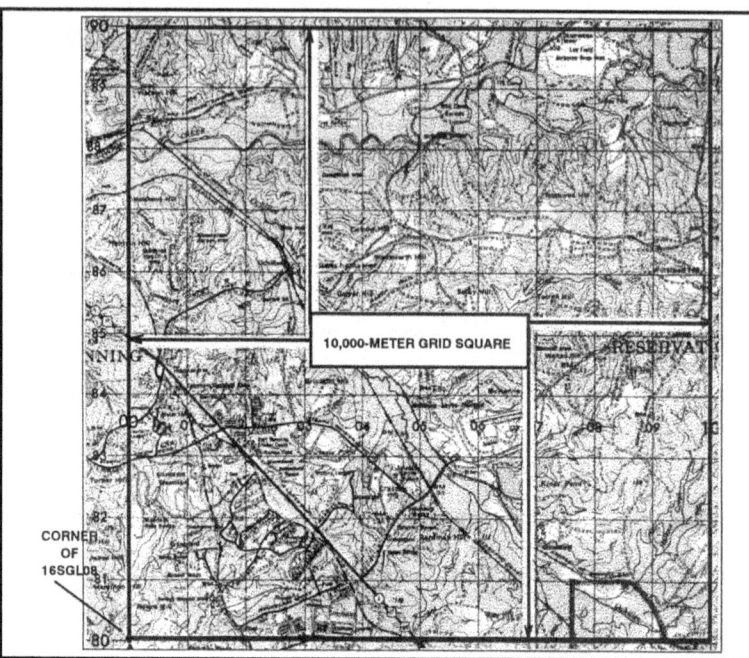

Figure 4-18. The 10,000-meter grid square.

e. **1,000-Meter Square**. To obtain 1,000-meter squares, each side of the meter square is divided into 10 equal parts. This division appears on large-scale n the actual grid lines; they are 1,000 meters apart. On the Columbus map. coordinates 16SGL0182, the easting 01 and the northing 82 gives the location southwest corner of grid square 0182 or to the nearest 1,000 meters of a point on t (Figure 4-19).

FM 3-25.26

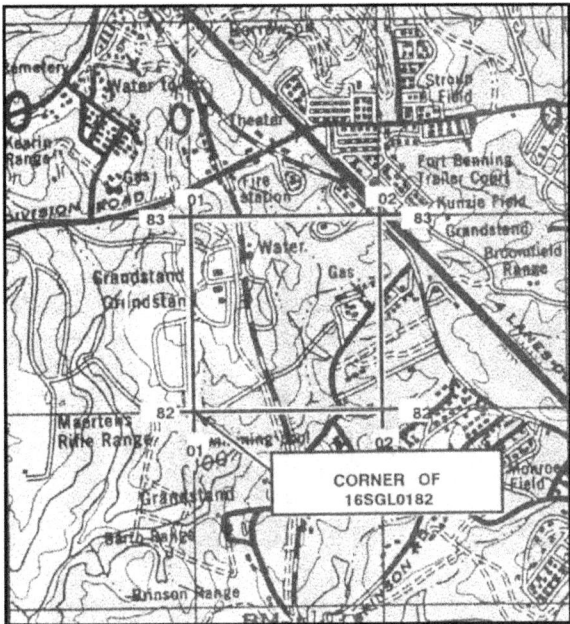

Figure 4-19. The 1,000-meter grid square.

f. **100-Meter Identification.** To locate to the nearest 100 meters, the grid coordinate scale can be used to divide the 1,000-meter grid squares into 10 equal parts (Figure 4-20, page 4-22).

g. **10-Meter Identification.** The grid coordinate scale has divisions every 50 meters on the 1:50,000 scale and every 20 meters on the 1:25,000 scale. These can be used to estimate to the nearest 10 meters and give the location of one point on the earth's surface to the nearest 10 meters. For example: 16SGL01948253 (gas tank).

FM 3-25.26

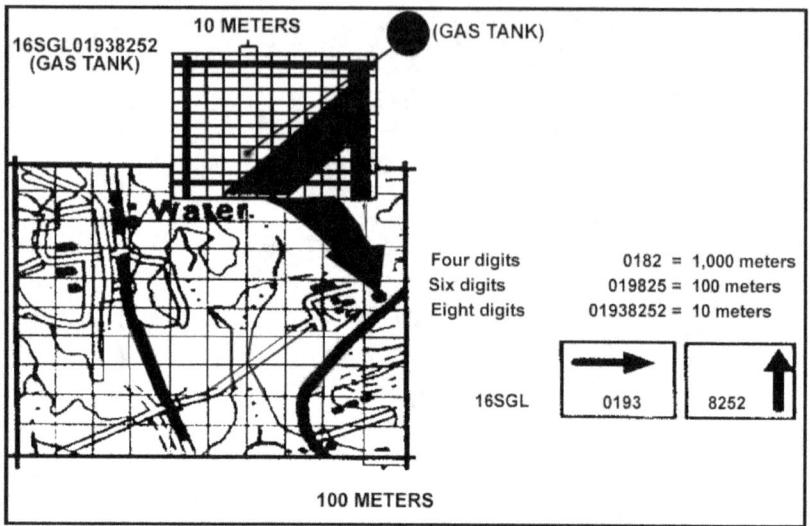

Figure 4-20. The 100-meter and 10-meter grid squares.

h. **Precision.** The precision of a point's location is shown by the number of digits in the coordinates; the more digits, the more precise the location (Figure 4-20, insert).

4-7. GRID REFERENCE BOX

A grid reference box (Figure 4-21) appears in the marginal information of each map sheet. It contains step-by-step instructions for using the grid and the U.S. Army military grid reference system. The grid reference box is divided into two parts.

a. The left portion identifies the grid zone designation and the 100,000-meter square. If the sheet falls in more than one 100,000-meter square, the grid lines that separate the squares are shown in the diagram and the letters identifying the 100,000-meter squares are given.

EXAMPLE: On the Columbus map sheet, the vertical line labeled 00 is the grid line that separates the two 100,000-meter squares, FL and GL. The left portion also shows a sample for the 1,000-meter square with its respective labeled grid coordinate numbers and a sample point within the 1,000-meter square.

b. The right portion of the grid reference box explains how to use the grid and is keyed on the sample 1,000-meter square of the left side. The following is an example of the military grid reference:

EXAMPLE: 16S locates the 6° by 8° area (grid zone designation).

4-22   FOUO   18 January 2005

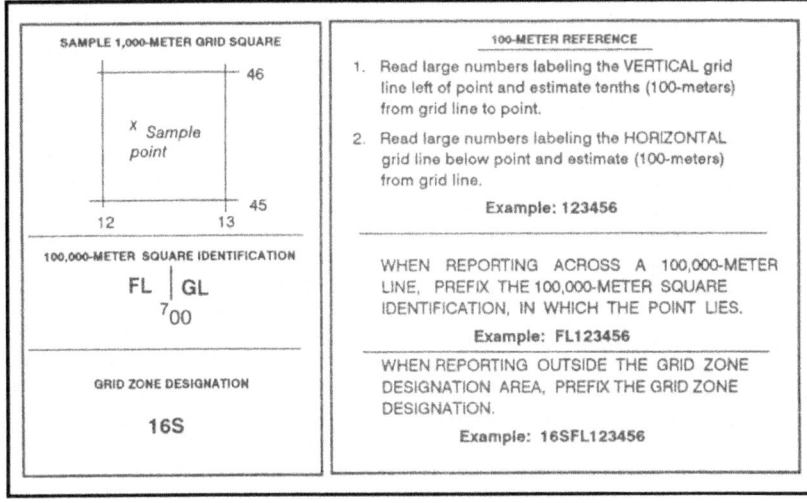

Figure 4-21. Grid reference box.

4-8. OTHER GRID SYSTEMS

The military grid reference system is not universally used. You must be prepared to interpret and use other grid systems, depending on your area of operations or the personnel you are operating with.

   a. **British Grids**. In a few areas of the world, British grids are still shown on military maps. However, the British grid systems are being phased out. Eventually all military mapping will be converted to the UTM grid.

   b. **World Geographic Reference System (GEOREF).** This is a worldwide position reference system used primarily by the U.S. Air Force. It may be used with any map or chart that has latitude and longitude printed on it. Instructions for using GEOREF data are printed in blue and are found in the margin of aeronautical charts (Figure 4-20, page 4-24). This system is based upon a division of the earth's surface into quadrangles of latitude and longitude having a systematic identification code. It is a method of expressing latitude and longitude in a form suitable for rapid reporting and plotting. Figure 4-20 illustrates a sample grid reference box using GEOREF. The GEOREF system uses an identification code that has three main divisions.

FM 3-25.26

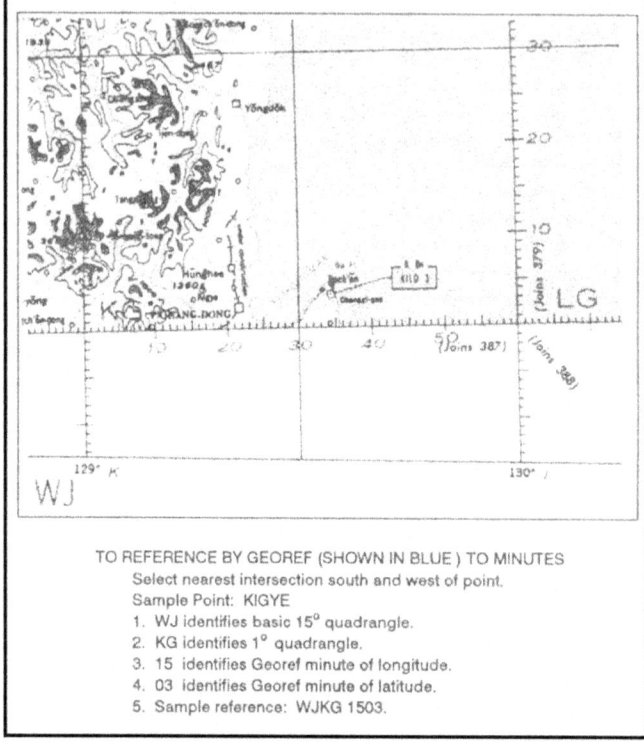

Figure 4-20. Sample reference using GEOREF.

(1) *First Division*. There are 24 north-south (longitudinal) zones, each 15° wide. These zones, starting at 180° and progressing eastward, are lettered A through Z (omitting I and O). The first letter of any GEOREF coordinate identifies the north-south zone in which the point is located. There are 12 east-west (latitudinal) bands, each 15° wide. These bands are lettered A through M (omitting I) northward from the south pole. The second letter of any GEOREF coordinate identifies the east-west band in which the point is located. The zones and bands divide the earth's surface into 288 quadrangles, each identified by two letters.

(2) *Second Division*. Each 15° quadrangle is further divided into 225 quadrangles of 1° each (15° by 15°). This division is effected by dividing a basic 15° quadrangle into 15 north-south zones and 15 east-west bands. The north-south zones are lettered A through Q (omitting I and O) from west to east. The third letter of any GEOREF coordinate identifies the 1° north-south zone within a 15° quadrangle. The east-west bands are lettered A through Q (I and O omitted) from south to north. The fourth letter of a GEOREF coordinate identifies the 1° east-west band within a 15° quadrangle. Four letters will identify any 1° quadrangle in the world.

(3) *Third Division*. Each of the 1° quadrangles is divided into 3,600 1" quadrangles. These 1" quadrangles are formed by dividing the 1° quadrangles into 60 1" north-south

4-24   FOUO
18 January 2005

zones numbered 0 through 59 from west to east, and 60 east-west bands numbered 0 to 59 from south to north. To designate any one of the 3,600 1" quadrangles requires four letters and four numbers. The rule READ RIGHT AND UP is always followed. Numbers 1 through 9 are written as 01, 02, and so forth. Each of the 1" quadrangles may be further divided into 10 smaller divisions both north-south and east-west, permitting the identification of 0.1" quadrangles. The GEOREF coordinate for any 0.1"quadrangle consists of four letters and six numbers.

4-9. PROTECTION OF MAP COORDINATES AND LOCATIONS

A disadvantage of any standard system of location is that the enemy, if he intercepts one of our messages using the system, can interpret the message and find our location. This possibility can be eliminated by using an authorized low-level numerical code to express locations. Army Regulation 380-40 outlines the procedures for obtaining authorized codes.

    a. The authorized numerical code provides a capability for encrypting map references and other numerical information that requires short-term security protection when, for operational reasons, the remainder of the message is transmitted in plain language. The system is published in easy-to-use booklets with sufficient material in each for one month's operation. Sample training editions of this type of system are available through the unit's communications and electronics officer.

    b. The use of any encryption methods other than authorized codes is, by regulation, unauthorized and shall not be used.

This Page intentionally left blank.

CHAPTER 5
SCALE AND DISTANCE

A map is a scaled graphic representation of a portion of the earth's surface. The scale of the map permits the user to convert distance on the map to distance on the ground or vice versa. The ability to determine distance on a map, as well as on the earth's surface, is an important factor in planning and executing military missions.

5-1. REPRESENTATIVE FRACTION

The numerical scale of a map indicates the relationship of distance measured on a map and the corresponding distance on the ground. This scale is usually written as a fraction and is called the representative fraction (RF). The RF is always written with the map distance as 1 and is independent of any unit of measure. (It could be yards, meters, inches, and so forth.) An RF of 1/50,000 or 1:50,000 means that one unit of measure on the map is equal to 50,000 units of the same measure on the ground.

    a. The ground distance between two points is determined by measuring between the same two points on the map and then multiplying the map measurement by the denominator of the RF or scale (Figure 5-1).

EXAMPLE:
    The map scale is 1:50,000
    RF = 1/50,000
    The map distance from point A to point B is 5 units
    5 x 50,000 = 250,000 units of ground distance

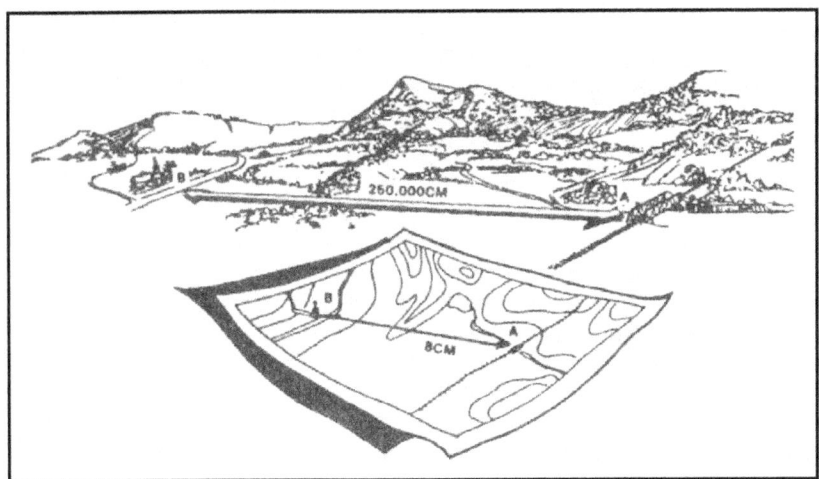

Figure 5-1. Converting map distance to ground distance.

b. Since the distance on most maps is marked in meters and the RF is expressed in this unit of measurement in most cases, a brief description of the metric system is needed. In the metric system, the standard unit of measurement is the meter.

    1 meter contains 100 centimeters (cm).
    100 meters is a regular football field plus 10 meters.
    1,000 meters is 1 kilometer (km).
    10,000 meters is 10 kilometers.

NOTE: Appendix C contains the units of measure conversion tables.

c. The situation may arise when a map or sketch has no RF or scale. To be able to determine ground distance on such a map, the RF must be determined. There are two ways to do this:

(1) *Comparison with Ground Distance.* Measure the distance between two points on the map—map distance (MD). Determine the horizontal distance between these same two points on the ground—ground distance (GD). Use the RF formula and remember that RF must be in the general form:

$$RF = \frac{1}{X} = \frac{MD}{GD}$$

Both the MD and the GD must be in the same unit of measure and the MD must be reduced to 1.

EXAMPLE:

    MD = 4.32 centimeters

    GD = 2.16 kilometers
    (216,000 centimeters)

$$RF = \frac{1}{X} = \frac{4.32}{216,000}$$

    or

$$\frac{216,000}{4.32} = 50,000$$

    therefore

$$RF = \frac{1}{50,000} \quad \text{or} \quad 1:50,000$$

(2) *Comparison With Another Map of the Same Area that Has an RF.* Select two points on the map with the unknown RF, and measure the distance (MD) between them. Locate those same two points on the map that has the known RF, and measure the distance (MD) between them. Using the RF for this map, determine GD, which is the same for both maps. Using the GD and the MD from the first map, determine the RF using the formula:

$$RF = \frac{1}{X} = \frac{MD}{GD}$$

Occasionally it may be necessary to determine map distance from a known ground distance and the RF:

$$MD = \frac{GD}{\text{Denominator or RF}}$$

Ground Distance = 2,200 meters

RF = 1:50,000

$$MD = \frac{2,200 \text{ meters}}{50,000}$$

MD = 0.044 meters x 100 (centimeters per meter)

MD = 4.4 centimeters

d. When determining ground distance from a map, the scale of the map affects the accuracy. As the scale becomes smaller, the accuracy of measurement decreases because some of the features on the map must be exaggerated so that they may be readily identified.

5-2. GRAPHIC (BAR) SCALES

A graphic scale is a ruler printed on the map and is used to convert distances on the map to actual ground distances. The graphic scale is divided into two parts. To the right of the zero, the scale is marked in full units of measure and is called the primary scale. To the left of the zero, the scale is divided into tenths and is called the extension scale. Most maps have three or more graphic scales, each using a different unit of measure (Figure 5-2, page 5-4). When using the graphic scale, be sure to use the correct scale for the unit of measure desired.

FM 3-25.26

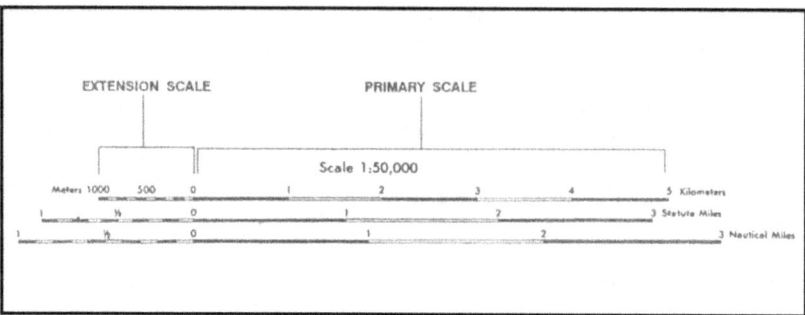

Figure 5-2. Using a graphic (bar) scale.

a. To determine straight-line distance between two points on a map, lay a straight-edged piece of paper on the map so that the edge of the paper touches both points and extends past them. Make a tick mark on the edge of the paper at each point (Figure 5-3).

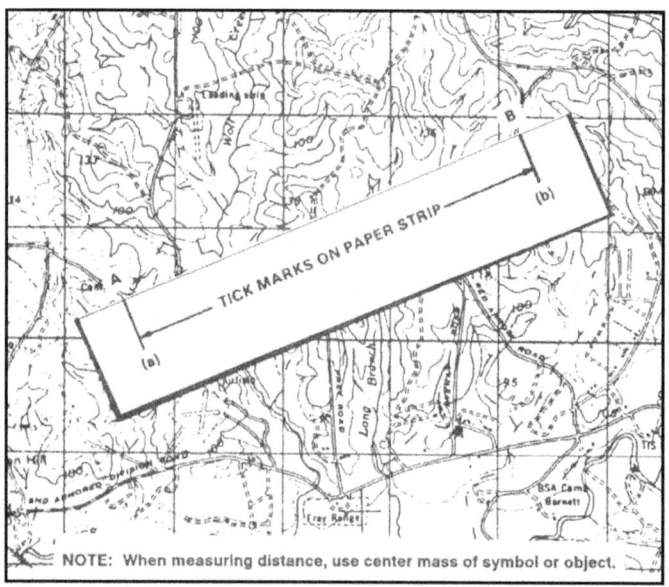

Figure 5-3. Transferring map distance to paper strip.

b. To convert the map distance to ground distance, move the paper down to the graphic bar scale, and align the right tick mark (b) with a printed number in the primary scale so that the left tick mark (a) is in the extension scale (Figure 5-4).

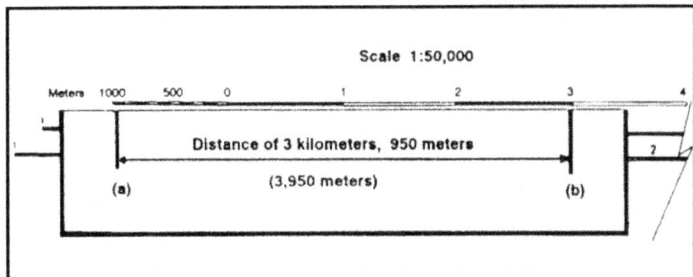

Figure 5-4. Measuring straight-line map distance.

c. The right tick mark (b) is aligned with the 3,000-meter mark in the primary scale, thus the distance is at least 3,000 meters. To determine the distance between the two points to the nearest 10 meters, look at the extension scale. The extension scale is numbered with zero at the right and increases to the left. When using the extension scale, always read right to left (Figure 5-4). From the zero left to the beginning of the first shaded area is 100 meters. From the beginning to the end of the shaded square is about 100 to 200 meters. From the end of the first shaded square to the beginning of the second shaded square is about 200 to 300 meters. Remember, the distance in the extension scale increases from right to left.

d. To determine the distance from the zero to tick mark (a), divide the distance inside the squares into tenths (Figure 5-4). As you break down the distance between the squares in the extension scale into tenths, you will see that tick mark (a) is aligned with the 950-meter mark. Adding the distance of 3,000 meters determined in the primary scale to the 950 meters determined by using the extension scale, the total distance between points (a) and (b) is 3,950 meters.

e. To measure distance along a road, stream, or other curved line, the straight edge of a piece of paper is used. In order to avoid confusion concerning the point to begin measuring from and the ending point, an eight-digit coordinate should be given for both the starting and ending points. Place a tick mark on the paper and map at the beginning point from which the curved line is to be measured. Align the edge of the paper along a straight portion and make a tick mark on both map and paper when the edge of the paper leaves the straight portion of the line being measured (A, Figure 5-5, page 5-6).

f. Keeping both tick marks together (on paper and map), place the point of the pencil close to the edge of the paper on the tick mark to hold it in place. Then, pivot the paper until another straight portion of the curved line is aligned with the edge of the paper. Continue in this manner until the measurement is completed (B, Figure 5-5, page 5-6).

g. When you have completed measuring the distance, move the paper to the graphic scale to determine the ground distance. The only tick marks you will be measuring the distance between are tick marks (a) and (b). The tick marks in between are not used (C, Figure 5-5, page 5-6).

FM 3-25.26

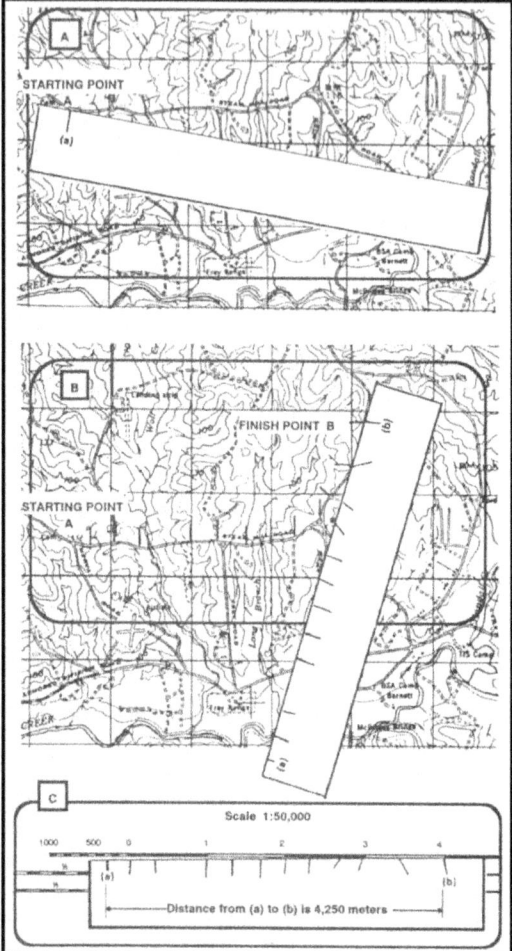

Figure 5-5. Measuring a curved line.

h. There may be times when the distance you measure on the edge of the paper exceeds the graphic scale. In this case, there are different techniques you can use to determine the distance.

(1) One technique is to align the right tick mark (b) with a printed number in the primary scale, in this case the 5. You can see that from point (a) to point (b) is more than 6,000 meters when you add the 1,000 meters in the extension scale. To determine the exact distance to the nearest 10 meters, place a tick mark (c) on the edge of the paper at the end of the extension scale (A, Figure 5-6). You know that from point (b) to point (c) is 6,000 meters. With the tick mark (c) placed on the edge of the paper at the end of the extension scale, slide the paper to the right. Remember the distance in the extension is always read from right to left. Align tick mark (c) with zero and then measure the distance between tick

marks (a) and (c). The distance between tick marks (a) and (c) is 420 meters. The total ground distance between start and finish points is 6,420 meters (B, Figure 5-6).

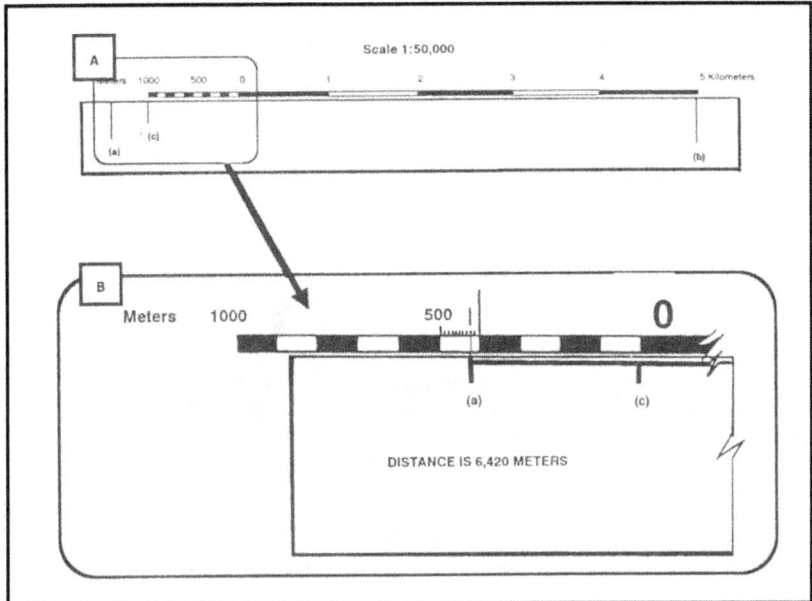

Figure 5-6. Determining the exact distance.

(2) Another technique that may be used to determine exact distance between two points when the edge of the paper exceeds the bar scale is to slide the edge of the paper to the right until tick mark (a) is aligned with the edge of the extension scale. Make a tick mark on the paper, in line with the 2,000-meter mark (c) (A, Figure 5-7, page 5-8). Then slide the edge of the paper to the left until tick mark (b) is aligned with the zero. Estimate the 100-meter increments into 10-meter increments to determine how many meters tick mark (c) is from the zero line (B, Figure 5-7, page 5-8). The total distance would be 3,030 meters.

(3) At times you may want to know the distance from a point on the map to a point off the map. In order to do this, measure the distance from the start point to the edge of the map. The marginal notes give the road distance from the edge of the map to some towns, highways, or junctions off the map. To determine the total distance, add the distance measured on the map to the distance given in the marginal notes. Be sure the unit of measure is the same.

(4) When measuring distance in statute or nautical miles, round it off to the nearest one-tenth of a mile and make sure the appropriate bar scale is used.

(5) Distance measured on a map does not take into consideration the rise and fall of the land. All distances measured by using the map and graphic scales are flat distances. Therefore, the distance measured on a map will increase when actually measured on the ground. This must be taken into consideration when navigating across country.

FM 3-25.26

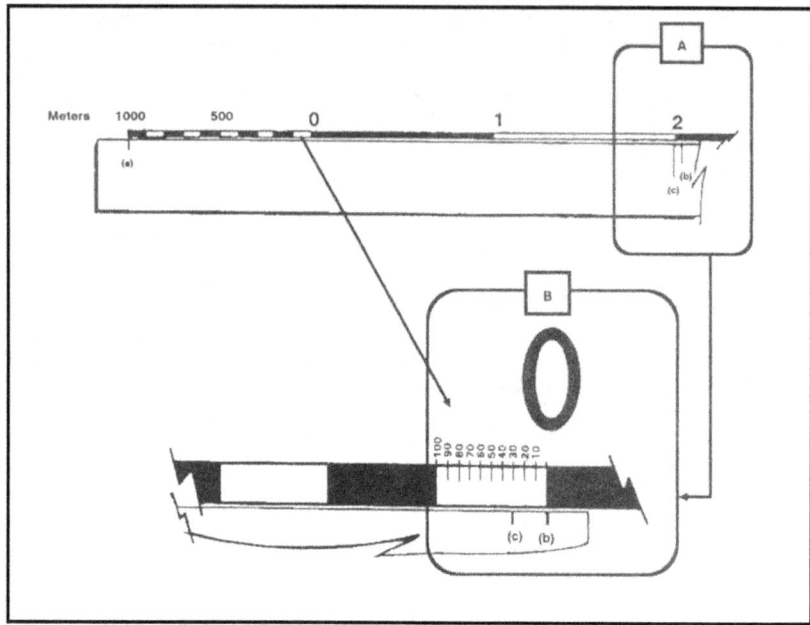

Figure 5-7. Reading the extension scale.

i. The amount of time required to travel a certain distance on the ground is an important factor in most military operations. This can be determined if a map of the area is available and a graphic time-distance scale is constructed for use with the map as follows:

R = Rate of travel (speed)
T = Time
D = Distance (ground distance)

$$T = \frac{D}{R}$$

For example, if an infantry unit is marching at an average rate (R) of 4 kilometers per hour, it will take about 3 hours (T) to travel 12 kilometers (D).

$$\frac{12\ (D)}{4\ (R)} = 3\ (T)$$

j. To construct a time-distance scale (A, Figure 5-8), knowing your length of march, rate of speed, and map scale (that is, 12 kilometers at 3 kilometers per hour on a 1:50,000-scale map), use the following process:

(1) Mark off the total distance on a line by referring to the graphic scale of the map or, if this is impracticable, compute the length of the line as follows:

(a) Convert the ground distance to centimeters: 12 kilometers x 100,000 (centimeters per kilometer) = 1,200,000 centimeters.

(b) Find the length of the line to represent the distance at map scale.

$$MD = \frac{1}{50,000} = \frac{1,200,000}{50,000} = 24 \text{ centimeters}$$

(c) Construct a line 24 centimeters in length (A, Figure 5-8).

(2) Divide the line by the rate of march into three parts (B, Figure 5-8), each part representing the distance traveled in one hour, and label.

(3) Divide the scale extension (left portion) into the desired number of lesser time divisions.

1-minute divisions — 60
5-minute divisions — 12
10-minute divisions — 6

(4) C, Figure 5-8 shows a 5-minute interval scale. Make these divisions in the same manner as for a graphic scale. The completed scale makes it possible to determine where the unit will be at any given time. However, it must be remembered that this scale is for one specific rate of march only, 4 kilometers per hour.

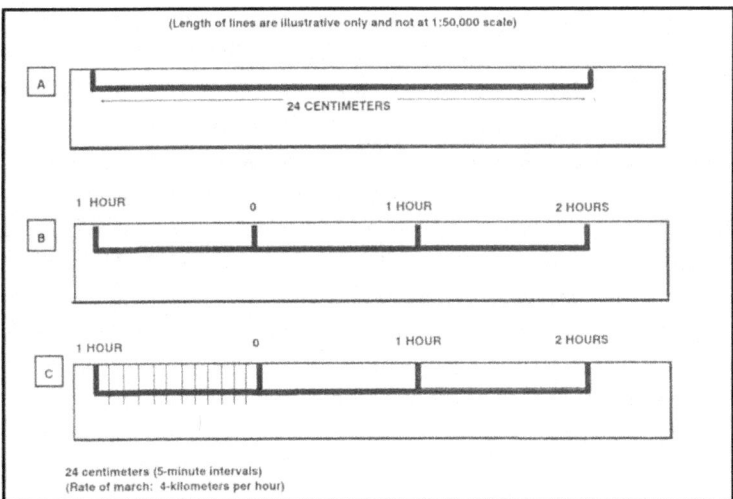

Figure 5-8. Constructing a time-distance scale.

5-3. OTHER METHODS

Determining distance is the most common source of error encountered while moving either mounted or dismounted. There may be circumstances where you are unable to determine distance using your map or where you are without a map. It is, therefore, essential to learn methods by which you can accurately pace, measure, use subtense, or estimate distances on the ground.

a. **Pace Count.** Another way to measure ground distance is the pace count. A pace is equal to one natural step, about 30 inches long. To accurately use the pace count method, you must know how many paces it takes you to walk 100 meters. To determine this, you must walk an accurately measured course and count the number of paces you take. A pace course can be as short as 100 meters or as long as 600 meters. The pace course, regardless of length, must be on similar terrain to that you will be walking over. It does no good to walk a course on flat terrain and then try to use that pace count on hilly terrain. To determine your pace count on a 600-meter course, count the paces it takes you to walk the 600 meters, then divide the total paces by 6. The answer will give you the average paces it takes you to walk 100 meters. It is important that each person who navigates while dismounted knows his pace count.

(1) There are many methods to keep track of the distance traveled when using the pace count. Some of these methods are: put a pebble in your pocket every time you have walked 100 meters according to your pace count; tie knots in a string; or put marks in a notebook. Do not try to remember the count; always use one of these methods or design your own method.

(2) Certain conditions affect your pace count in the field, and you must allow for them by making adjustments.

(a) *Slopes.* Your pace lengthens on a downslope and shortens on an upgrade. Keeping this in mind, if it normally takes you 120 paces to walk 100 meters, your pace count may increase to 130 or more when walking up a slope.

(b) *Winds.* A head wind shortens the pace and a tail wind increases it.

(c) *Surfaces.* Sand, gravel, mud, snow, and similar surface materials tend to shorten the pace.

(d) *Elements.* Falling snow, rain, or ice cause the pace to be reduced in length.

(e) *Clothing.* Excess clothing and boots with poor traction affect the pace length.

(f) *Visibility.* Poor visibility, such as in fog, rain, or darkness, will shorten your pace.

b. **Odometer.** Distances can be measured by an odometer, which is standard equipment on most vehicles. Readings are recorded at the start and end of a course and the difference is the length of the course.

(1) To convert kilometers to miles, multiply the number of kilometers by 0.62.

EXAMPLE:
16 kilometers = 16 x 0.62 = 9.92 miles

(2) To convert miles to kilometers, divided the number of miles by 0.62.

EXAMPLE:
10 miles = 10 divided by 0.62 = 16.12 kilometers

c. **Subtense.** The subtense method is a fast method of determining distance and yields accuracy equivalent to that obtained by measuring distance with a premeasured piece of wire. An advantage is that a horizontal distance is obtained indirectly; that is, the distance is computed rather than measured. This allows subtense to be used over terrain where obstacles, such as streams, ravines, or steep slopes, may prohibit other methods of determining distance.

(1) The principle used in determining distance by the subtense method is similar to that used in estimating distance by the mil relation formula. The field artillery (FA) application of the mil relation formula involves only estimations. It is not accurate enough for survey purposes. However, the subtense method uses precise values with a trigonometric solution. Subtense is based on a principle of visual perspective—the farther away an object is, the smaller it appears.

(2) The following two procedures are involved in subtense measurement:
- Establishing a base of known length.
- Measuring the angle of that base by use of the aiming circle.

(3) The subtense base may be any desired length. However, if a 60-meter base, a 2-meter bar, or the length of an M16A1 or M16A2 rifle is used, precomputed subtense tables are available. The M16 or 2-meter bar must be held horizontal and perpendicular to the line of sight by a soldier facing the aiming circle. The instrument operator sights on one end of the M16 or 2-meter bar and measures the horizontal clockwise angle to the other end of the rifle or bar. He does this twice and averages the angles. He then enters the appropriate subtense table with the mean angle and extracts the distance. Accurate distances can be obtained with the M16 out to approximately 150 meters, with the 2-meter bar out to 250 meters, and with the 60-meter base out to 1,000 meters. If a base of another length is desired, a distance can be computed by using the following formula:

$$\text{Distance} = \frac{1/2 \text{ (base in meters)}}{\text{Tan } (1/2) \text{ (in mils)}}$$

d. **Estimation.** At times, because of the tactical situation, it may be necessary to estimate range. There are two methods that may be used to estimate range or distance.

(1) *100-Meter Unit-of-Measure Method.* To use this method, the soldier must be able to visualize a distance of 100 meters on the ground. For ranges up to 500 meters, he determines the number of 100-meter increments between the two objects he wishes to measure. Beyond 500 meters, the soldier must select a point halfway to the object(s) and determine the number of 100-meter increments to the halfway point, then double it to find the range to the object (Figure 5-9, page 5-12).

Figure 5-9. Using a 100-meter unit-of-measure method.

(2) *Flash-to-Bang Method.* To use this method to determine range to an explosion or enemy fire, begin to count when you see the flash. Count the seconds until you hear the weapon fire. This time interval may be measured with a stopwatch or by using a steady count, such as one-thousand-one, one-thousand-two, and so forth, for a three-second estimated count. If you must count higher than 10 seconds, start over with one. Multiply the number of seconds by 330 meters to get the approximate range (FA uses 350 meters instead).

(3) *Proficiency of Methods.* The methods discussed above are used only to estimate range (Table 5-1). Proficiency in both methods requires constant practice. The best training technique is to require the soldier to pace the range after he has estimated the distance. In this way, the soldier discovers the actual range for himself, which makes a greater impression than if he is simply told the correct range.

Factors Affecting Range Estimation	Factors Causing Underestimation of Range	Factors Causing Overestimation of Range
The clearness of outline and details of the object.	When most of the object is visible and offers a clear outline.	When only a small part of the object can be seen or the object is small in relation to its surroundings.
Nature of terrain or position of the observer.	When looking across a depression that is mostly hidden from view.	When looking across a depression that is totally visible.
	When looking downward from high ground.	When vision is confined, as in streets, draws, or forest trails.
	When looking down a straight, open road or along a railroad.	When looking from low ground toward high ground.
	When looking over uniform surfaces like water, snow, desert, or grain fields.	In poor light, such as dawn and dusk; in rain, snow, fog; or when the sun is in the observer's eyes.
	In bright light or when the sun is shining from behind the observer.	
Light and atmosphere	When the object is in sharp contrast with the background or is silhouetted because of its size, shape, or color.	When object blends into the background or terrain.
	When seen in the clear air of high altitudes.	

Table 5-1. Factors of range estimation.

This Page intentionally left blank.

CHAPTER 6
DIRECTION

Being in the right place at the prescribed time is necessary to successfully accomplish military missions. Direction plays an important role in a soldier's everyday life. It can be expressed as right, left, straight ahead, and so forth; but then the question arises, "To the right of what?" This chapter defines the word azimuth and the three different norths. It explains in detail how to determine the grid and the magnetic azimuths with the use of the protractor and the compass. It explains the use of some field-expedient methods to find directions, the declination diagram, and the conversion of azimuths from grid to magnetic and vice versa. It also includes some advanced aspects of map reading such as intersection, resection, modified resection, and polar plots.

6-1. METHODS OF EXPRESSING DIRECTION

Military personnel need a way of expressing direction that is accurate, is adaptable to any part of the world, and has a common unit of measure. Directions are expressed as units of angular measure.

    a. **Degree.** The most common unit of measure is the degree (°) with its subdivisions of minutes (') and seconds (").

    1 degree = 60 minutes.
    1 minute = 60 seconds.

    b. **Mil.** Another unit of measure, the mil (abbreviated ḿ in graphics), is used mainly in artillery, tank, and mortar gunnery. The mil expresses the size of an angle formed when a circle is divided into 6,400 angles, with the vertex of the angles at the center of the circle. A relationship can be established between degrees and mils. A circle equals 6400 mils divided by 360 degrees, or 17.78 mils per degree. To convert degrees to mils, multiply degrees by 17.78.

    c. **Grad.** The grad is a metric unit of measure found on some foreign maps. There are 400 grads in a circle (a 90-degree right angle equals 100 grads). The grad is divided into 100 centesimal minutes (centigrads) and the minute into 100 centesimal seconds (milligrads).

6-2. BASE LINES

In order to measure something, there must always be a starting point or zero measurement. To express direction as a unit of angular measure, there must be a starting point or zero measure and a point of reference. These two points designate the base or reference line. There are three base lines—true north, magnetic north, and grid north (Figure 6-1, page 6-2). The most commonly used are magnetic and grid.

    a. **True North.** True north is defined as a line from any point on the earth's surface to the north pole. All lines of longitude are true north lines. True north is usually represented by a star.

FM 3-25.26

b. **Magnetic North**. Magnetic north is the direction to the north magnetic pole, as indicated by the north-seeking needle of a magnetic instrument. The magnetic north is usually symbolized by a line ending with half of an arrowhead. Magnetic readings are obtained with magnetic instruments such as lensatic and M2 compasses.

c. **Grid North**. Grid north is the north that is established by using the vertical grid lines on the map. Grid north may be symbolized by the letters GN or the letter "y".

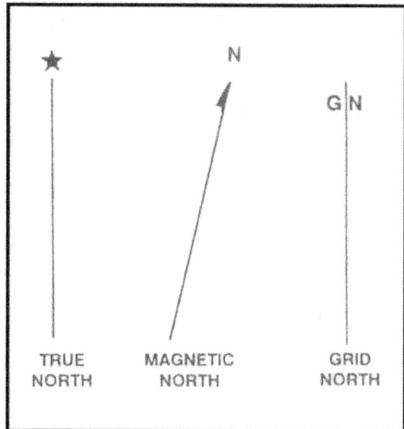

Figure 6-1. Three norths.

6-3. AZIMUTHS

An azimuth is defined as a horizontal angle measured clockwise from a north base line. This north base line could be true north, magnetic north, or grid north. The azimuth is the most common military method to express direction. When using an azimuth, the point from which the azimuth originates is the center of an imaginary circle (Figure 6-2). This circle is divided into 360 degrees or 6400 mils.

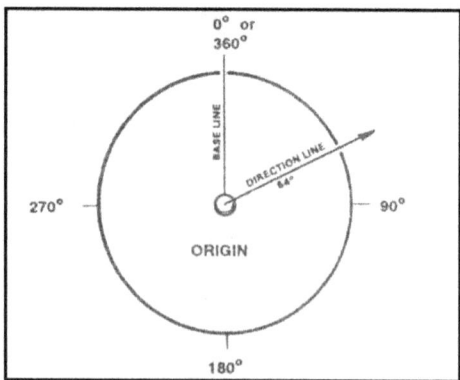

Figure 6-2. Origin of azimuth circle.

6-2   FOUO
18 January 2005

a. **Back Azimuth.** A back azimuth is the opposite direction of an azimuth. It is comparable to doing "about face." To obtain a back azimuth from an azimuth, add 180 degrees if the azimuth is 180 degrees or less; subtract 180 degrees if the azimuth is 180 degrees or more (Figure 6-3). The back azimuth of 180 degrees may be stated as 0 degrees or 360 degrees. For mils, if the azimuth is less than 3200 mils, add 3200 mils; if the azimuth is more than 3200 mils, subtract 3200 mils.

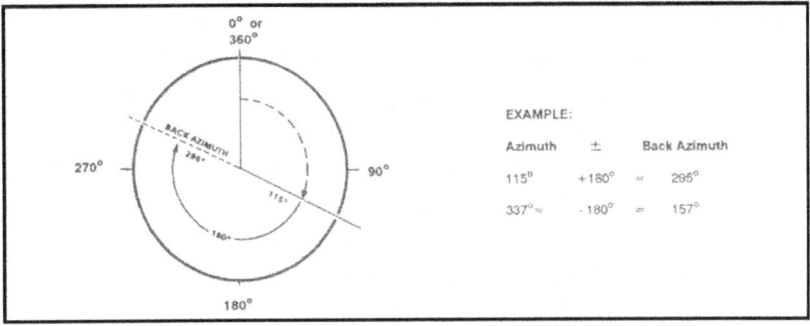

Figure 6-3. Back azimuth.

WARNING
When converting azimuths into back azimuths, extreme care should be exercised when adding or subtracting the 180 degrees. A simple mathematical mistake could cause disastrous consequences.

b. **Magnetic Azimuth.** The magnetic azimuth is determined by using magnetic instruments such as lensatic and M2 compasses. (See Chapter 9 for details.)

c. **Field-Expedient Methods.** Several field-expedient methods to determine direction are discussed in Chapter 9.

6-4. GRID AZIMUTHS

When an azimuth is plotted on a map between point A (starting point) and point B (ending point), the points are joined together by a straight line. A protractor is used to measure the angle between grid north and the drawn line, and this measured azimuth is the grid azimuth (Figure 6-4, page 6-4).

FM 3-25.26

> **WARNING**
> When measuring azimuths on a map, remember that you are measuring from a starting point to an ending point. If a mistake is made and the reading is taken from the ending point, the grid azimuth will be opposite, thus causing the user to go in the wrong direction.

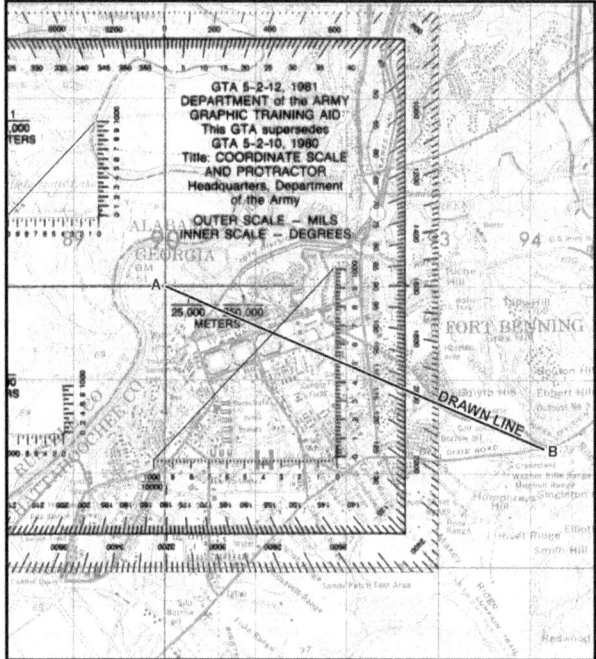

Figure 6-4. Measuring an azimuth.

6-5. PROTRACTOR

There are several types of protractors—full circle, half circle, square, and rectangular (Figure 6-5). All of them divide the circle into units of angular measure, and each has a scale around the outer edge and an index mark. The index mark is the center of the protractor circle from which all directions are measured.

6-4   FOUO
                      18 January 2005

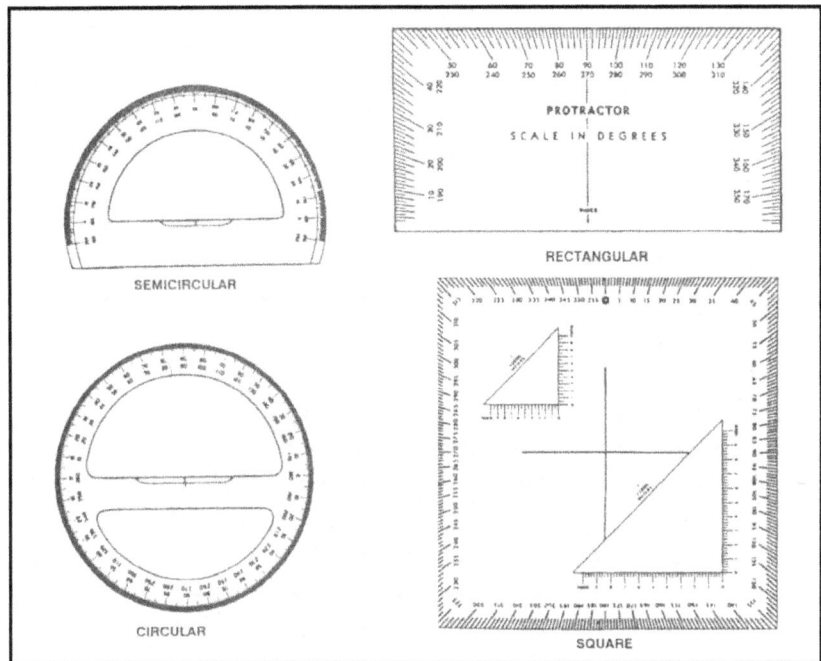

Figure 6-5. Types of protractors.

a. The military protractor, GTA 5-2-12, contains two scales: one in degrees (inner scale) and one in mils (outer scale). This protractor represents the azimuth circle. The degree scale is graduated from 0 to 360 degrees with each tick mark representing one degree. A line from 0 to 180 degrees is called the base line of the protractor. The index or center of the protractor is where the base line intersects the horizontal line, between 90 and 270 degrees (Figure 6-6, page 6-6).

FM 3-25.26

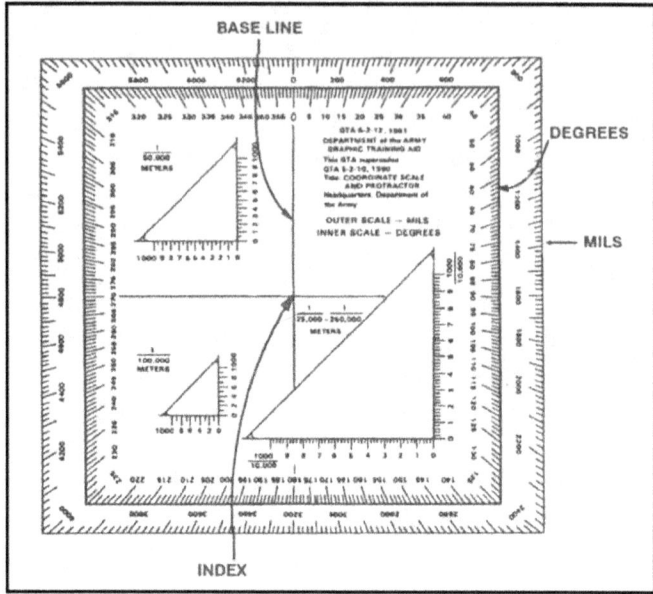

Figure 6-6. Military protractor.

b. When using the protractor, the base line is always oriented parallel to a north-south grid line. The 0- or 360-degree mark is always toward the top or north on the map and the 90-degree mark is to the right.

(1) To determine the grid azimuth—
(a) Draw a line connecting the two points (A and B).
(b) Place the index of the protractor at the point where the drawn line crosses a vertical (north-south) grid line.
(c) Keeping the index at this point, align the 0- to 180-degree line of the protractor on the vertical grid line.
(d) Read the value of the angle from the scale; this is the grid azimuth from point A to point B (Figure 6-4, page 6-5).

(2) To plot an azimuth from a known point on a map (Figure 6-7)—
(a) Convert the azimuth from magnetic to grid, if necessary (see paragraph 6-6).
(b) Place the protractor on the map with the index mark at the center of mass of the known point and the base line parallel to a north-south grid line.
(c) Make a mark on the map at the desired azimuth.
(d) Remove the protractor and draw a line connecting the known point and the mark on the map. This is the grid direction line (azimuth).

NOTE: When measuring an azimuth, the reading is always to the nearest degree or 10 mils. Distance does not change an accurately measured azimuth.

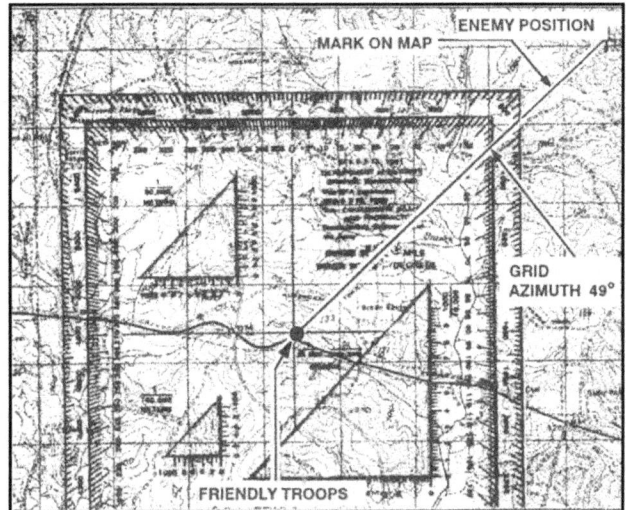

Figure 6-7. Plotting an azimuth on the map.

c. To obtain an accurate reading with the protractor (to the nearest degree or 10 mils), there are two techniques to check that the base line of the protractor is parallel to a north-south grid line.

(1) Place the protractor index where the azimuth line cuts a north-south grid line, aligning the base line of the protractor directly over the intersection of the azimuth line with the north-south grid line. The user should be able to determine whether the initial azimuth reading was correct.

(2) The user should re-read the azimuth between the azimuth and north-south grid line to check the initial azimuth.

(3) Note that the protractor is cut at both the top and bottom by the same north-south grid line. Count the number of degrees from the 0-degree mark at the top of the protractor to this north-south grid line and then count the number of degrees from the 180-degree mark at the bottom of the protractor to this same grid line. If the two counts are equal, the protractor is properly aligned.

6-6. DECLINATION DIAGRAM

Declination is the angular difference between any two norths. If you have a map and a compass, the declination of most interest to you will be between magnetic and grid north. The declination diagram (Figure 6-8, page 6-8) shows the angular relationship, represented by prongs, among grid, magnetic, and true norths. While the relative positions of the prongs are correct, they are seldom plotted to scale. Do not use the diagram to measure a numerical value. This value will be written in the map margin (in both degrees and mils) beside the diagram.

a. **Location.** A declination diagram is a part of the information in the lower margin on most larger maps. On medium-scale maps, the declination information is shown by a note in the map margin.

b. **Grid-Magnetic Angle.** The G-M angle value is the angular size that exists between grid north and magnetic north. It is an arc, indicated by a dashed line, that connects the grid-north and magnetic-north prongs. This value is expressed to the nearest 1/2 degree, with mil equivalents shown to the nearest 10 mils. The G-M angle is important to the map reader/land navigator because azimuths translated between map and ground will be in error by the size of the declination angle if not adjusted for it.

c. **Grid Convergence.** An arc indicated by a dashed line connects the prongs for true north and grid north. The value of the angle for the center of the sheet is given to the nearest full minute with its equivalent to the nearest mil. These data are shown in the form of a grid-convergence note.

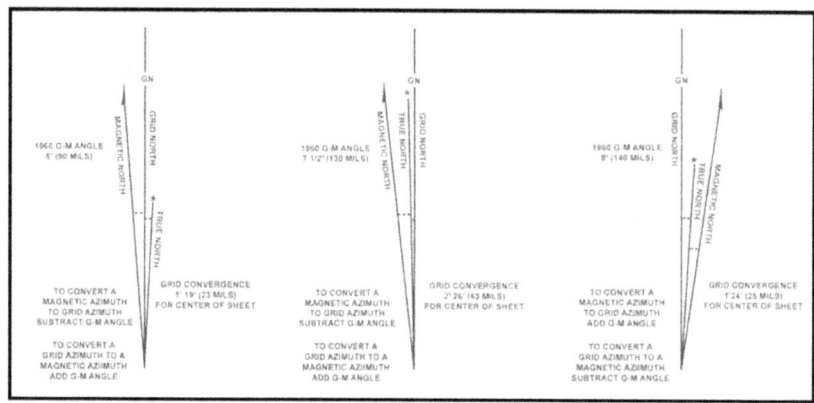

Figure 6-8. Declination diagrams.

d. **Conversion.** There is an angular difference between the grid north and the magnetic north. Since the location of magnetic north does not correspond exactly with the grid-north lines on the maps, a conversion from magnetic to grid or vice versa is needed.

(1) *With Notes.* Simply refer to the conversion notes that appear in conjunction with the diagram explaining the use of the G-M angle (Figure 6-8). One note provides instructions for converting magnetic azimuth to grid azimuth; the other, for converting grid azimuth to magnetic azimuth. The conversion (add or subtract) is governed by the direction of the magnetic-north prong relative to that of the grid-north prong.

(2) *Without Notes.* In some cases, there are no declination conversion notes on the margin of the map; it is necessary to convert from one type of declination to another. A magnetic compass gives a magnetic azimuth; but in order to plot this line on a gridded map, the magnetic azimuth value must be changed to grid azimuth. The declination diagram is used for these conversions. A rule to remember when solving such problems is: **No matter where the azimuth line points, the angle to it is always measured clockwise from the reference direction (base line).** With this in mind, the problem is solved using the following steps:

(a) Draw a vertical or grid-north line (prong). Always align this line with the vertical lines on a map (Figure 6-9).

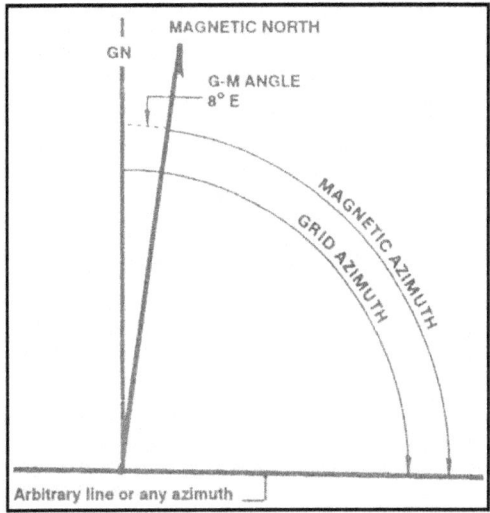

Figure 6-9. Declination diagram with arbitrary line.

(b) From the base of the grid-north line (prong), draw an arbitrary line (or any azimuth line) at a roughly right angle to north, regardless of the actual value of the azimuth in degrees (Figure 6-9).

(c) Examine the declination diagram on the map and determine the direction of the magnetic north (right-left or east-west) relative to that of the grid-north prong. Draw a magnetic prong from the apex of the grid-north line in the desired direction (Figure 6-9).

(d) Determine the value of the G-M angle. Draw an arc from the grid prong to the magnetic prong and place the value of the G-M angle (Figure 6-9).

(e) Complete the diagram by drawing an arc from each reference line to the arbitrary line. A glance at the completed diagram shows whether the given azimuth or the desired azimuth is greater, and, thus, whether the known difference between the two must be added or subtracted.

(f) The inclusion of the true-north prong in relationship to the conversion is of little importance.

e. **Applications.** Remember, there are no negative azimuths on the azimuth circle. Since 0 degree is the same as 360 degrees, then 2 degrees is the same as 362 degrees. This is because 2 degrees and 362 degrees are located at the same point on the azimuth circle. The grid azimuth can now be converted into a magnetic azimuth because the grid azimuth is now larger than the G-M angle.

(1) When working with a map having an east G-M angle:
(a) To plot a magnetic azimuth on a map, first change it to a grid azimuth (Figure 6-10).

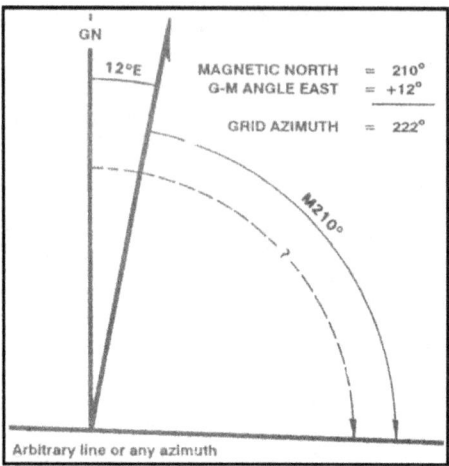

Figure 6-10. Converting to grid azimuth.

(b) To use a magnetic azimuth in the field with a compass, first change the grid azimuth plotted on a map to a magnetic azimuth (Figure 6-11).

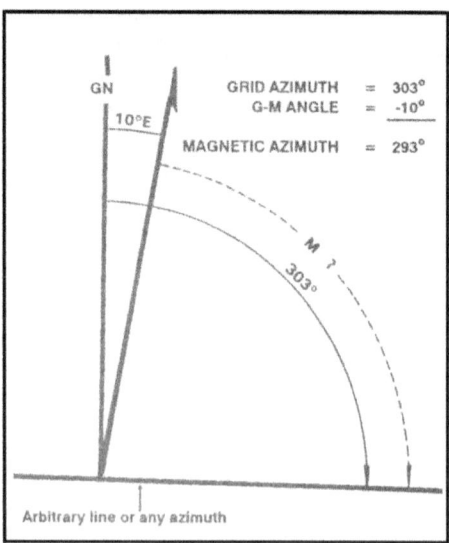

Figure 6-11. Converting to magnetic azimuth.

(c) Convert a grid azimuth to a magnetic azimuth when the G-M angle is greater than a grid azimuth (Figure 6-12).

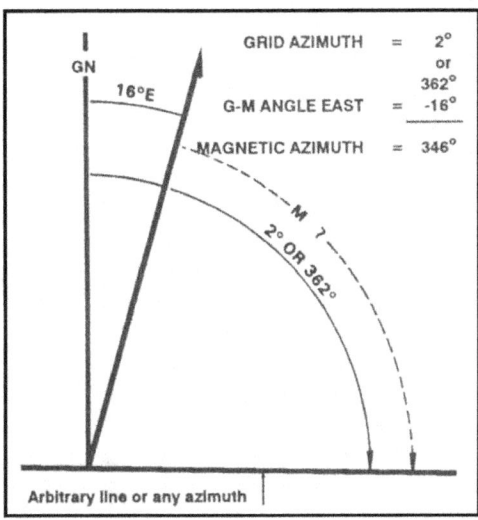

Figure 6-12. Converting to a magnetic azimuth when the G-M angle is greater.

(2) When working with a map having a west G-M angle:
(a) To plot a magnetic azimuth on a map, first convert it to a grid azimuth (Figure 6-13).

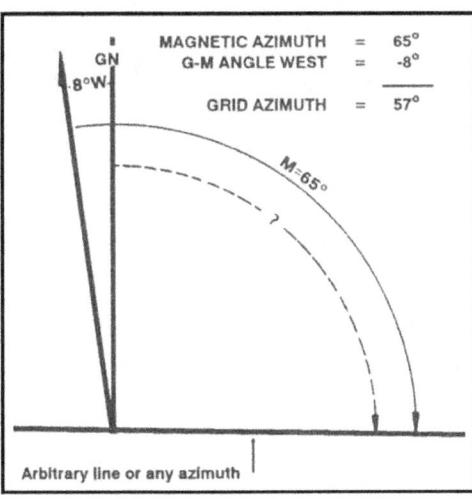

Figure 6-13. Converting to a grid azimuth on a map.

(b) To use a magnetic azimuth in the field with a compass, change the grid azimuth plotted on a map to a magnetic azimuth (Figure 6-14).

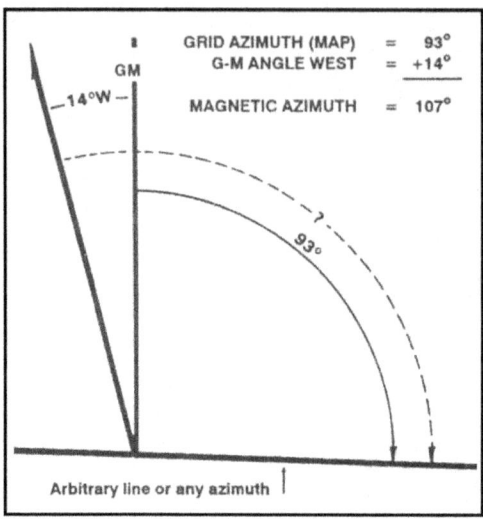

Figure 6-14. Converting to a magnetic azimuth on a map.

(c) Convert a magnetic azimuth when the G-M angle is greater than the magnetic azimuth (Figure 6-15).

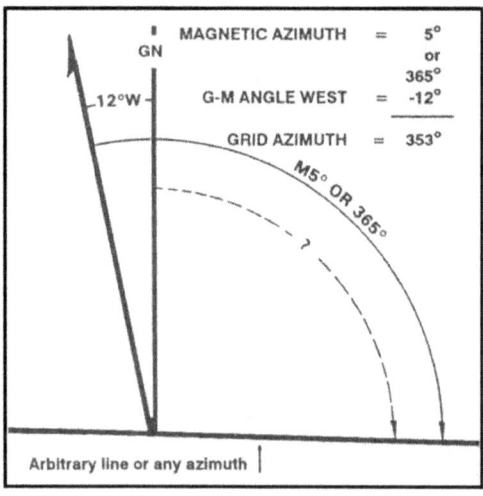

Figure 6-15. Converting to a grid azimuth when the G-M angle is greater.

(3) The G-M angle diagram should be constructed and used each time the conversion of azimuth is required. Such procedure is important when working with a map for the first time. It also may be convenient to construct a G-M angle conversion table on the margin of the map.

NOTE: When converting azimuths, exercise extreme care when adding and subtracting the G-M angle. A simple mistake of 1 degree could be significant in the field.

6-7. INTERSECTION

Intersection is the location of an unknown point by successively occupying at least two (preferably three) known positions on the ground and then map sighting on the unknown location. It is used to locate distant or inaccessible points or objects such as enemy targets and danger areas. There are two methods of intersection: the map and compass method and the straightedge method (Figures 6-16 and 6-17 on page 6-14).

a. When using the map and compass method—
(1) Orient the map using the compass.
(2) Locate and mark your position on the map.
(3) Determine the magnetic azimuth to the unknown position using the compass.
(4) Convert the magnetic azimuth to grid azimuth.
(5) Draw a line on the map from your position on this grid azimuth.
(6) Move to a second known point and repeat steps 1, 2, 3, 4, and 5.
(7) The location of the unknown position is where the lines cross on the map. Determine the grid coordinates to the desired accuracy.

b. The straightedge method is used when a compass is not available. When using it—
(1) Orient the map on a flat surface by the terrain association method.
(2) Locate and mark your position on the map.
(3) Lay a straightedge on the map with one end at the user's position (A) as a pivot point; then, rotate the straightedge until the unknown point is sighted along the edge.
(4) Draw a line along the straightedge
(5) Repeat the above steps at position (B) and check for accuracy.

FM 3-25.26

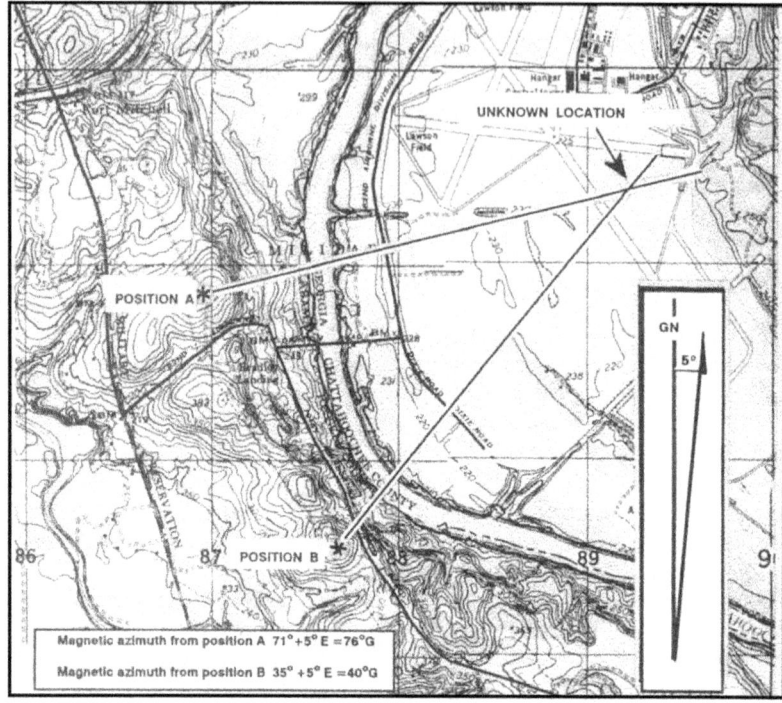

Figure 6-16. Intersection, using map and compass.

(6) The intersection of the lines on the map is the location of the unknown point (C). Determine the grid coordinates to the desired accuracy (Figure 6-17).

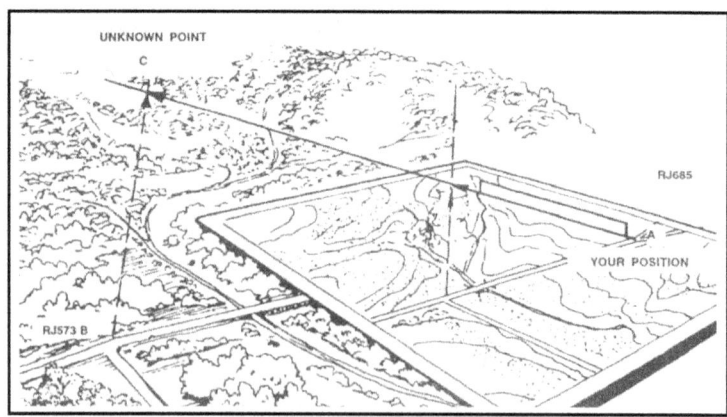

Figure 6-17. Intersection, using a straightedge.

6-8. RESECTION

Resection is the method of locating one's position on a map by determining the grid azimuth to at least two well-defined locations that can be pinpointed on the map. For greater accuracy, the desired method of resection would be to use three or more well-defined locations.

a. When using the map and compass method (Figure 6-18)—

(1) Orient the map using the compass.

(2) Identify two or three known distant locations on the ground and mark them on the map.

(3) Measure the magnetic azimuth to one of the known positions from your location using a compass.

(4) Convert the magnetic azimuth to a grid azimuth.

(5) Convert the grid azimuth to a back azimuth. Using a protractor, draw a line for the back azimuth on the map from the known position back toward your unknown position.

(6) Repeat 3, 4, and 5 for a second position and a third position, if desired.

(7) The intersection of the lines is your location. Determine the grid coordinates to the desired accuracy.

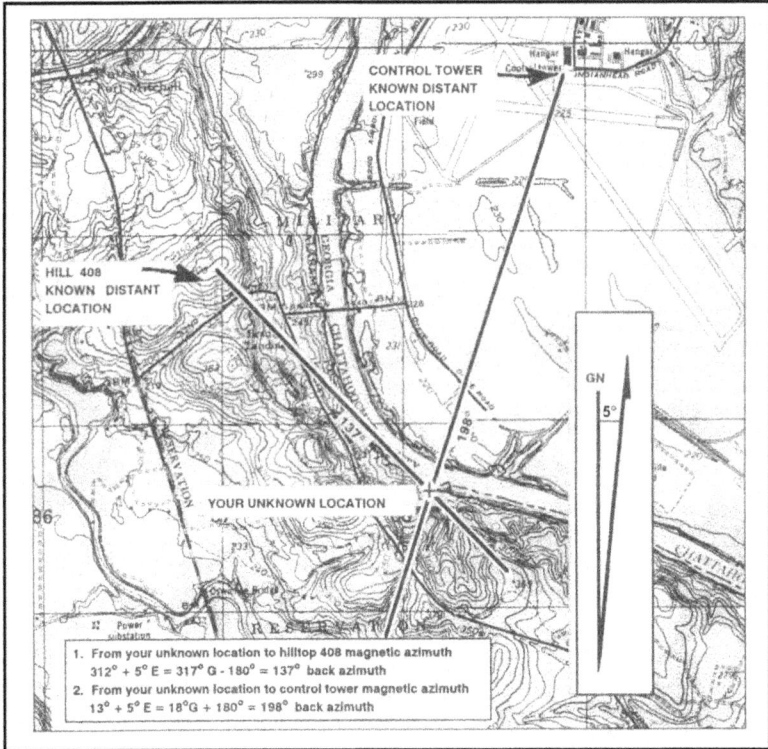

Figure 6-18. Resection with map and compass.

a. When using the straightedge method (Figure 6-19)—
(1) Orient the map on a flat surface by the terrain association method.
(2) Locate at least two known distant locations or prominent features on the ground and mark them on the map.
(3) Lay a straightedge on the map using a known position as a pivot point. Rotate the straightedge until the known position on the map is aligned with the known position on the ground.
(4) Draw a line along the straightedge away from the known position on the ground toward your position.
(5) Repeat 3 and 4 using a second known position.
(6) The intersection of the lines on the map is your location. Determine the grid coordinates to the desired accuracy.

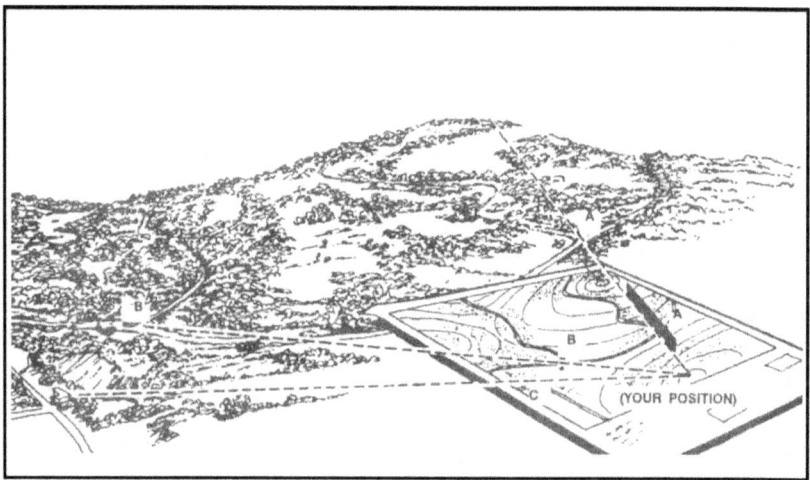

Figure 6-19. Resection with straightedge.

6-9. MODIFIED RESECTION

Modified resection is the method of locating one's position on the map when the person is located on a linear feature on the ground, such as a road, canal, or stream (Figure 6-20). Proceed as follows:
  a. Orient the map using a compass or by terrain association.
  b. Find a distant point that can be identified on the ground and on the map.
  c. Determine the magnetic azimuth from your location to the distant known point.
  d. Convert the magnetic azimuth to a grid azimuth.
  e. Convert the grid azimuth to a back azimuth. Using a protractor, draw a line for the back azimuth on the map from the known position back toward your unknown position.
  f. The location of the user is where the line crosses the linear feature. Determine the grid coordinates to the desired accuracy.

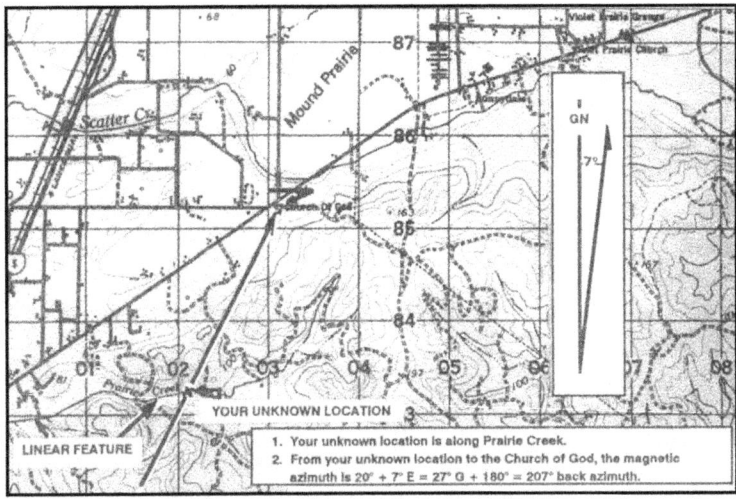

Figure 6-20. Modified resection.

6-10. POLAR PLOT

A method of locating or plotting an unknown position from a known point by giving a direction and a distance along that direction line is called polar plot. The following elements must be present when using polar plot (Figure 6-21).

- Present known location on the map.
- Azimuth (grid or magnetic).
- Distance (in meters).

Using the laser range finder to determine the range enhances your accuracy in determining the unknown position's location.

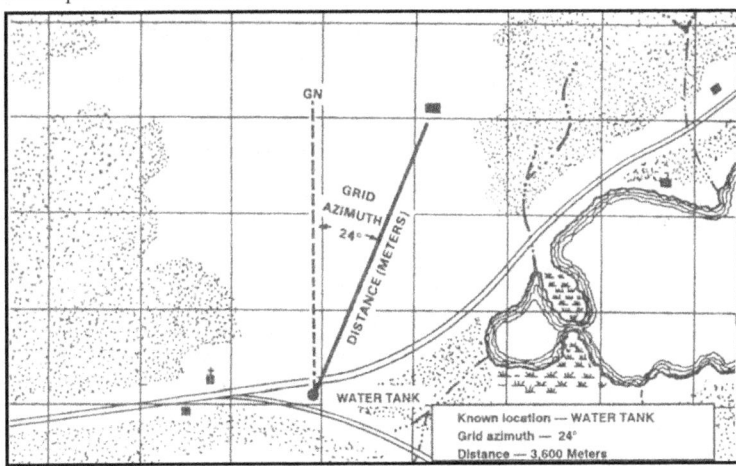

Figure 6-21. Polar plot.

This Page intentionally left blank.

FM 3-25.26

CHAPTER 7
OVERLAYS

An overlay is a clear sheet of plastic or semi-transparent paper. It is used to display supplemental map and tactical information related to military operations. It is often used as a supplement to orders given in the field. Information is plotted on the overlay at the same scale as on the map, aerial photograph, or other graphic being used. When the overlay is placed over the graphic, the details plotted on the overlay are shown in their true position.

7-1. PURPOSE
Overlays are used to display military operations with enemy and friendly troop dispositions, and as supplements to orders sent to the field. They show detail that will aid in understanding the orders, displays of communication networks, and so forth. They are also used as annexes to reports made in the field because they can clarify matters that are difficult to explain clearly in writing.

7-2. MAP OVERLAY
The three steps in making a map overlay are: orienting the overlay material, plotting and symbolizing the detail, and adding the required marginal information (Figure 7-1).

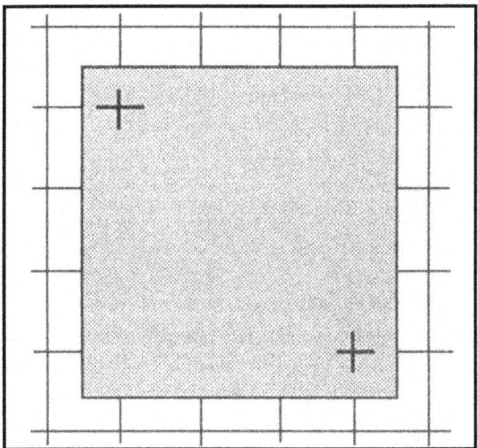

Figure 7-1. Orienting the overlay.

  a. **Orienting.** Orient the overlay over the place on the map to be annotated. Then, if possible, attach it to the edges of the map with tape. Trace the grid intersections nearest the two opposite corners of the overlay using a straightedge, and label each with the proper grid coordinates. These register marks show exactly where the overlay fits on the map; without them, the overlay is difficult to orient. It is imperative that absolute accuracy be maintained in plotting the register marks, as the smallest mistake will throw off the overlay.

FOUO

b. **Plotting New Detail.** To plot any detail, use pencils or markers in standard colors that make a lasting mark without cutting the overlay (FM 101-5-1).

(1) Use standard topographic or military symbols where possible. Nonstandard symbols invented by the author must be identified in a legend on the overlay. Depending on the conditions under which the overlay is made, it may be advisable to plot the positions first on the map, then trace them onto the overlay. Since the overlay is to be used as a supplement to orders or reports and the recipient will have an identical map, show only that detail with which the report is directly concerned.

(2) If you have observed any topographic or cultural features that are not shown on the map, such as a new road or a destroyed bridge, plot their positions as accurately as possible on the overlay and mark with the standard topographic symbol.

(3) If difficulty in seeing through the overlay material is encountered while plotting or tracing detail, lift the overlay from time to time to check orientation of information being added in reference to the base.

c. **Recording Marginal Information.** When all required detail has been plotted or traced on the overlay, print information as close to the lower right-hand corner as detail permits (Figure 7-2). This information includes the following data:

(1) *Title and Objective.* This tells the reader why the overlay was made and may also give the actual location. For example, "Road reconnaissance" is not as specific as "Route 146 road reconnaissance."

(2) *Time and Date.* Any overlay should contain the latest possible information. An overlay received in time is valuable to the planning staff and may affect the entire situation; an overlay that has been delayed for any reason may be of little use. Therefore, the exact time the information was obtained aids the receivers in determining its reliability and usefulness.

(3) *Map Reference.* The sheet name, sheet number, map series number, and scale must be included. If the reader does not have the same map that was used for the overlay, this provides the information necessary to obtain it.

(4) *Author.* The name, rank, and organization of the author, supplemented with a date and time of preparation of the overlay, tells the reader if there was a time difference between when the information was obtained and when it was reported.

(5) *Legend.* If it is necessary to invent nonstandard symbols to show the required information, the legend must show what these symbols mean.

(6) *Security Classification.* This must correspond to the highest classification of either the map or the information placed on the overlay. This will also be stated if the information and map are unclassified. The locations of the classification notes are shown in Figure 7-2, and the notes will appear in both locations as shown.

(7) *Additional Information.* Any other information that amplifies the overlay will also be included. Make it as brief as possible.

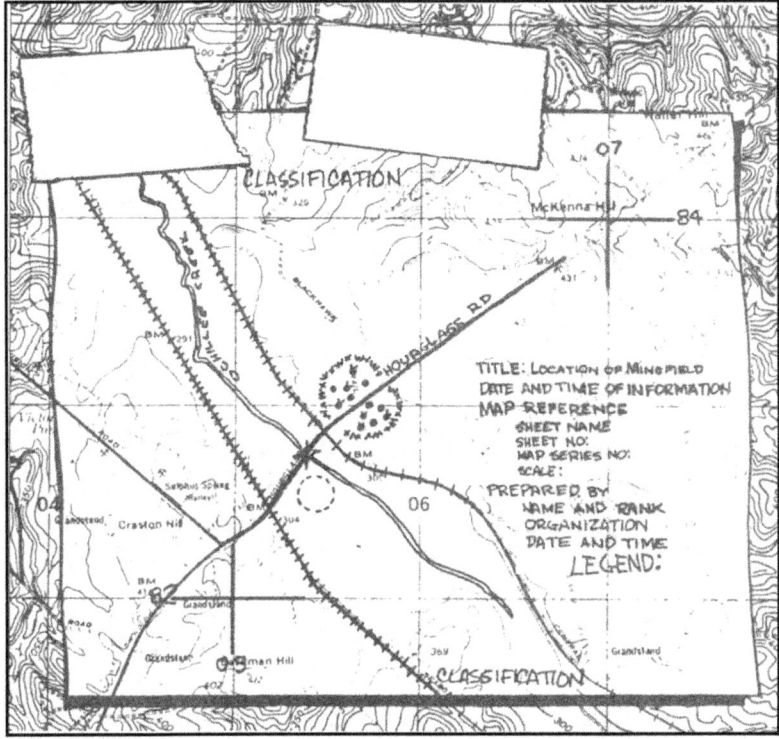

Figure 7-2. Map overlay with marginal information.

7-3. AERIAL PHOTOGRAPH OVERLAY

Overlays of single aerial photographs are constructed and used in the same way as map overlays. The steps followed are essentially the same, with the following exceptions:

a. **Orienting.** The photograph normally does not have grid lines to be used as register marks. The borders of the photograph limit the area of the overlay, so the reference marks or linear features are traced in place of grid register marks. Finally, to ensure proper location of the overlay with respect to the photograph, indicate on the overlay the position of the marginal data on the photograph as seen through the overlay.

b. **Recording Marginal Information.** The marginal information shown on photographs varies somewhat from that shown on maps. Overlays of photographs (Figure 7-3, page 7-4) should show the following information:

(1) *North Arrow.* This may be obtained in two ways—by comparing with a map of the area or by orienting the photograph by inspection. In the latter case, a compass or expedient direction finder must be used to place the direction arrow on the overlay. Use the standard symbol to represent the actual north arrow used—grid, magnetic, or true north.

(2) *Title and Objective.* This tells the reader why the photo overlay was made and may also give the actual location.

(3) **Time and Date.** The exact time the information was obtained is shown on a photo overlay just as on a map overlay

(4) **Photo Reference.** The photo number, mission number, date of flight, and scale appear here, or the information is traced in its actual location on the photograph.

(5) **Scale.** The scale must be computed since it is not part of the marginal data.

(6) **Map Reference.** Reference is made to the sheet name, sheet number, series number, and scale of a map of the area, if one is available.

(7) **Author.** The name, rank, and organization of the author are shown, supplemented with a date and time of preparation of the overlay.

(8) **Legend.** As with map overlays, this is only necessary when nonstandard symbols are used.

(9) **Security Classification.** This must correspond to the highest classification of either the photograph or the information placed on the overlay. It will also be stated if the information and photograph are unclassified. The locations of the classification notes are shown in Figure 7-3, and the notes will appear in both locations.

(10) **Additional Information.** Any other information that amplifies the overlay will also be included. Make it as brief as possible.

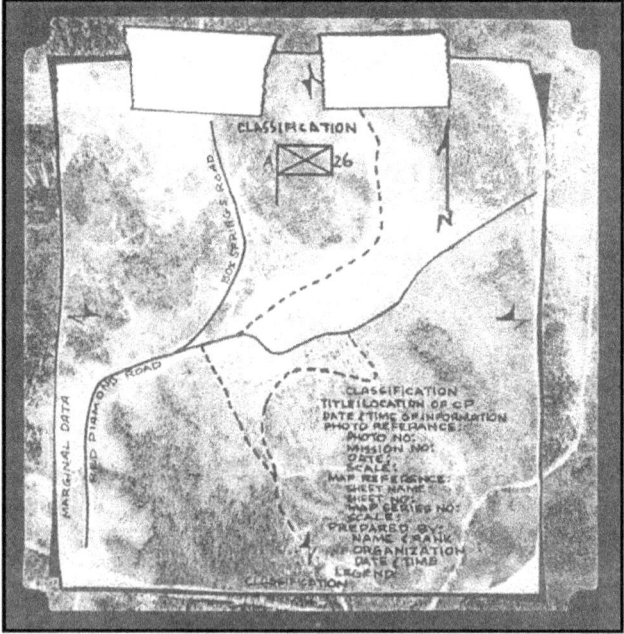

Figure 7-3. Photographic overlay with marginal information.

CHAPTER 8
AERIAL PHOTOGRAPHS

An aerial photograph is any photograph taken from an airborne vehicle (aircraft, drones, balloons, satellites, and so forth). The aerial photograph has many uses in military operations; however, for the purpose of this manual, it will be considered primarily as a map supplement or map substitute.

8-1. COMPARISON WITH MAPS

A topographic map may be obsolete because it was compiled many years ago. A recent aerial photograph shows any changes that have taken place since the map was made. For this reason, maps and aerial photographs complement each other. More information can be gained by using the two together than by using either alone.

    a. **Advantages.** An aerial photograph has the following advantages over a map:
- It provides a current pictorial view of the ground that no map can equal.
- It is more readily obtained. The photograph may be in the hands of the user within a few hours after it is taken; a map may take months to prepare.
- It may be made for places that are inaccessible to ground soldiers.
- It shows military features that do not appear on maps.
- It can provide a day-to-day comparison of selected areas, permitting evaluations to be made of enemy activity.
- It provides a permanent and objective record of the day-to-day changes with the area.

    b. **Disadvantages.** The aerial photograph has the following disadvantages as compared to a map:
- Ground features are difficult to identify or interpret without symbols and are often obscured by other ground detail such as, for example, buildings in wooded areas.
- Position location and scale are only approximate.
- Detailed variations in the terrain features are not readily apparent without overlapping photography and a stereoscopic viewing instrument.
- Because of a lack of contrasting colors and tone, a photograph is difficult to use in poor light.
- It lacks marginal data.
- It requires more training to interpret than a map.

8-2. TYPES

Aerial photography most commonly used by military personnel may be divided into two major types, the vertical and the oblique. Each type depends upon the attitude of the camera with respect to the earth's surface when the photograph is taken.

    a. **Vertical.** A vertical photograph is taken with the camera pointed as straight down as possible (Figure 8-1, page 8-2 and Figure 8-2, page 8-3). Allowable tolerance is usually + 3 degrees from the perpendicular (plumb) line to the camera axis. The result is coincident with the camera axis. A vertical photograph has the following characteristics:

- The lens axis is perpendicular to the surface of the earth.
- It covers a relatively small area.
- The shape of the ground area covered on a single vertical photo closely approximates a square or rectangle.
- Being a view from above, it gives an unfamiliar view of the ground.
- Distance and directions may approach the accuracy of maps if taken over flat terrain.
- Relief is not readily apparent.

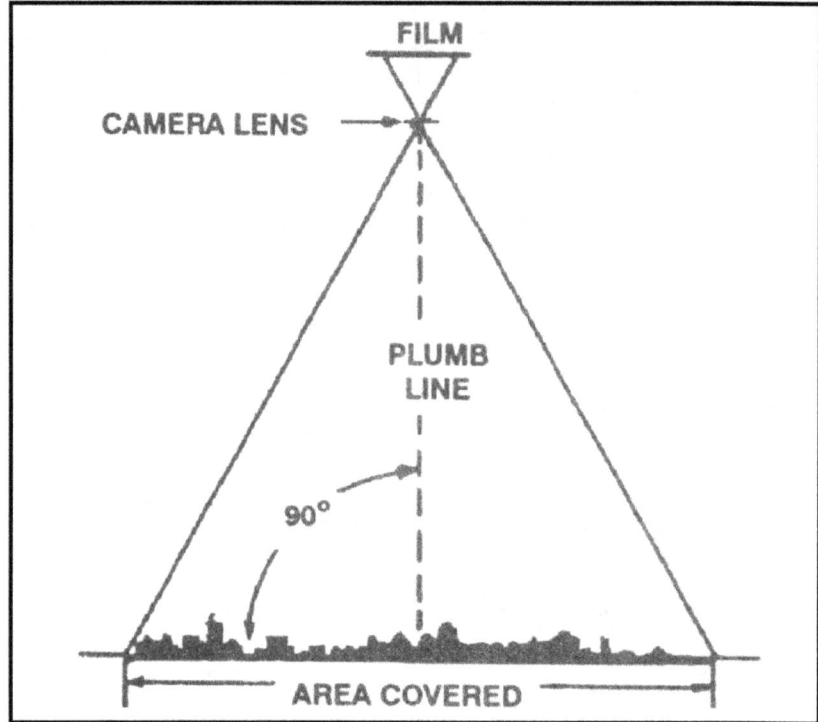

Figure 8-1. Relationship of the vertical aerial photograph with the ground.

Figure 8-2. Vertical photograph.

a. **Low Oblique.** A photograph taken with the camera inclined about 30 degrees from the vertical is a low oblique (Figures 8-3 and 8-4, page 8-4). It is used to study an area before an attack, to substitute for a reconnaissance, to substitute for a map, or to supplement a map. A low oblique has the following characteristics:
- It covers a relatively small area.
- The ground area covered is a trapezoid, although the photo is square or rectangular.
- The objects have a more familiar view, comparable to viewing from the top of a high hill or tall building.
- No scale is applicable to the entire photograph, and distance cannot be measured. Parallel lines on the ground are not parallel on this photograph; therefore, direction (azimuth) cannot be measured.
- Relief is discernible but distorted.
- It does not show the horizon.

FM 3-25.26

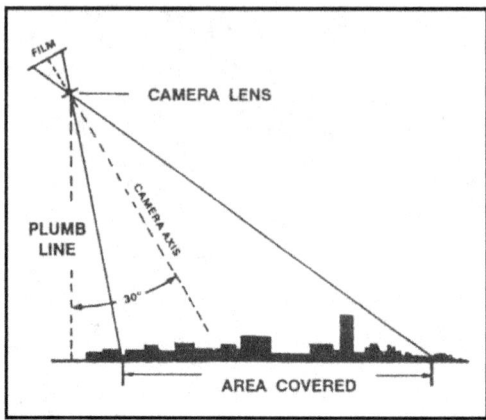

Figure 8-3. Relationship of low oblique photograph to the ground.

Figure 8-4. Low oblique photograph.

c. **High Oblique.** The high oblique is a photograph taken with the camera inclined about 60 degrees from the vertical (Figures 8-5 and 8-6). It is used primarily in the making of aeronautical charts and has a limited military application. However, it may be the only photography available. A high oblique has the following characteristics:
- It covers a very large area (not all usable).
- The ground area covered is a trapezoid, but the photograph is square or rectangular.
- The view varies from the very familiar to unfamiliar, depending on the height at which the photograph is taken.
- Distances and directions are not measured on this photograph for the same reasons that they are not measured on the low oblique.
- Relief may be quite discernible but distorted as in any oblique view. The relief is not apparent in a high altitude, high oblique.
- The horizon is always visible.

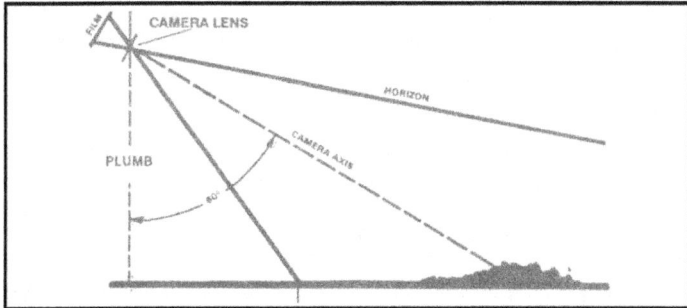

Figure 8-5. Relationship of high oblique photograph to the ground.

Figure 8-6. High oblique photograph.

d. **Trimetrogon.** A trimetrogon is an assemblage of three photographs taken at the same time, one vertical and two high obliques, in a direction at a right angle to the line of flight. The obliques, taken at an angle of 60 degrees from the vertical, overlap the sides of the vertical photography, producing composites from horizon to horizon (Figure 8-7).

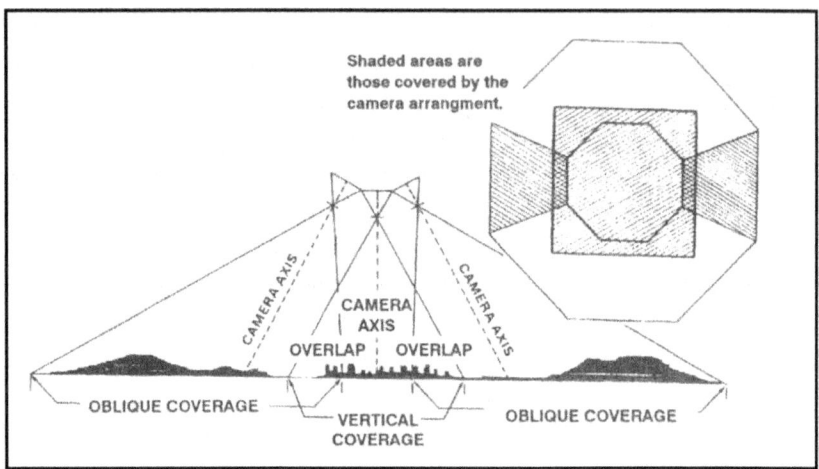

Figure 8-7. Relationship of cameras to ground for trimetrogon photography (three cameras).

e. **Multiple Lens Photography.** These are composite photographs taken with one camera having two or more lenses, or by two or more cameras. The photographs are combinations of two, four, or eight obliques around a vertical. The obliques are rectified to permit assembly as verticals on a common plane.

f. **Convergent Photography.** These are done with a single twin-lens, wide-angle camera, or with two single-lens, wide-angle cameras coupled rigidly in the same mount so that each camera axis converges when intentionally tilted a prescribed amount (usually 15 or 20 degrees) from the vertical. Again, the cameras are exposed at the same time. For precision mapping, the optical axes of the cameras are parallel to the line of flight, and for reconnaissance photography, the camera axes are at high angles to the line of flight.

g. **Panoramic.** The development and increasing use of panoramic photography in aerial reconnaissance has resulted from the need to cover in greater detail more and more areas of the world.

(1) To cover the large areas involved, and to resolve the desired ground detail, present-day reconnaissance systems must operate at extremely high-resolution levels. Unfortunately, high-resolution levels and wide-angular coverage are basically contradicting requirements.

(2) A panoramic camera is a scanning type of camera that sweeps the terrain of interest from side to side across the direction of flight. This permits the panoramic camera to record a much wider area of ground than either frame or strip cameras. As in the case of the frame cameras, continuous cover is obtained by properly spaced exposures timed to give sufficient

overlap between frames. Panoramic cameras are most advantageous for applications requiring the resolution of small ground detail from high altitudes.

8-3. TYPES OF FILM

Types of film generally used in aerial photography include panchromatic, infrared, and color. Camouflage detection film is also available.

   a. **Panchromatic.** This film is the same type of film that is used in a small, commercial hand-held camera. It records the amount of light reflected from objects in tones of gray running from white to black. Most aerial photography is taken with panchromatic film.

   b. **Infrared.** This is a black-and-white film that is sensitive to infrared waves. It can be used to detect artificial camouflage materials and to take photographs at night if there is a source of infrared radiation.

   c. **Color.** This film is the same as that used in an average, commercial hand-held camera. It is limited in its use because of the time required to process it and its need for clear, sunny weather.

   d. **Camouflage Detection.** This is a special type film that records natural vegetation in a reddish color. When artificial camouflage materials are photographed, they appear bluish or purplish. The name of this film indicates its primary use.

8-4. NUMBERING AND TITLING INFORMATION

Each aerial photograph contains in its margin important information for the photo user. The arrangement, type, and amount of this information is standardized; however, the rapid development of cameras, film, and aeronautical technology since World War II has caused numerous changes in the numbering and titling of aerial photographs. As a result, the photo user may find that the marginal information on older photographs varies somewhat from the standard current practice. With certain camera systems, some of the data are automatically recorded on each exposure, while other systems require that all titling data be added to the film after processing.

   a. Standard titling data for aerial photography prepared for the use of the Department of Defense are: for reconnaissance and charting photography, items 2 through 14 and item 19 are lettered on the beginning and end of each roll of film, and items 1 through 9 and item 19 are lettered on each exposure; for surveying and mapping photography, items 2 through 19 are lettered on the beginning and end of each roll of film, and items 1, 2, 3, 5, 6, 7, 8, 9, 13, and 19 are lettered on each exposure.

   (1) Negative number.
   (2) Camera position.
   (3) Taking unit.
   (4) Service.
   (5) Sortie/mission number.
   (6) Date (followed by a double hyphen [=]).
   (7) Time group and zone letter (GMT).
   (8) Focal length.
   (9) Altitude.
   (10) Kind of photography or imagery.
   (11) Geographic coordinates.
   (12) Descriptive title.

(13) Project number and or name.
(14) Camera type and serial number.
(15) Cone serial number (if any).
(16) Lens type and serial number.
(17) Magazine type and serial number.
(18) Type of photographic filter used.
(19) Security classification.

　　b. Automatically recorded data may differ somewhat in arrangement from the sequence listed above, but the same information is available to the photo user. (A detailed explanation of the titling items and the codes used to indicate them is found in TM 5-243.)

8-5. SCALE DETERMINATION

Before a photograph can be used as a map supplement or substitute, it is necessary to know its scale. On a map, the scale is printed as a representative fraction that expresses the ratio of map distance to ground distance. For example:

$$RF = \frac{MD}{GD}$$

On a photograph, the scale is also expressed as a ratio, but is the ratio of the photo distance (PD) to ground distance. For example:

$$RF = \frac{PD}{GD}$$

The approximate scale or average scale (RF) of a vertical aerial photograph is determined by either of two methods: the comparison method or the focal length-flight altitude method.

　　a. **Comparison Method.** The scale of a vertical aerial photograph is determined by comparing the measured distance between two points on the photograph with the measured ground distance between the same two points.

$$\text{SCALE RF} = \frac{\text{Photo Distance}}{\text{Ground Distance}}$$

The ground distance is determined by the actual measurement on the ground or by the use of the scale on a map of the same area. The points selected on the photograph must be identifiable on the ground or map of the same area and should be spaced in such a manner that a line connecting them will pass through or nearly through the center of the photograph (Figure 8-8).

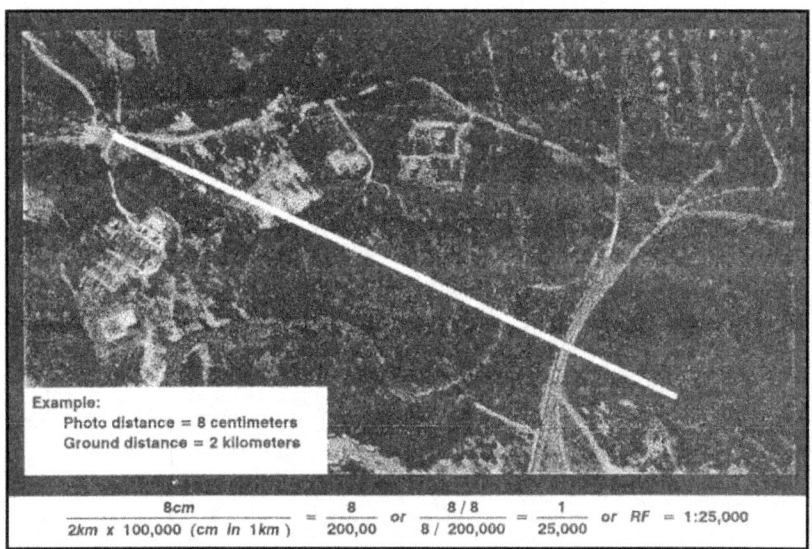

Figure 8-8. Selection of points for scale determination.

b. **Focal Length-Flight Altitude Method.** When the marginal information of a photograph includes the focal length and the flight altitude, the scale of the photo is determined using the following formula (Figure 8-9).

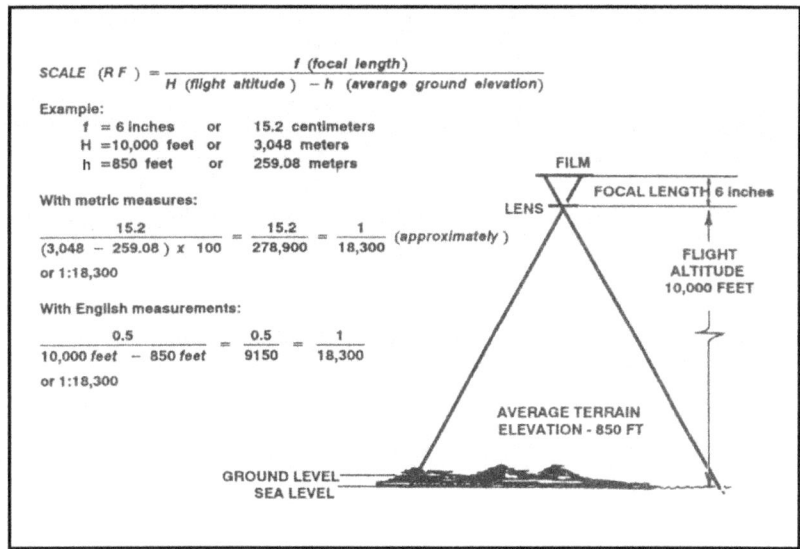

Figure 8-9. Computation of scale from terrain level.

When the ground elevation is at sea level, "h" becomes zero, and the formula is as shown in Figure 8-10.

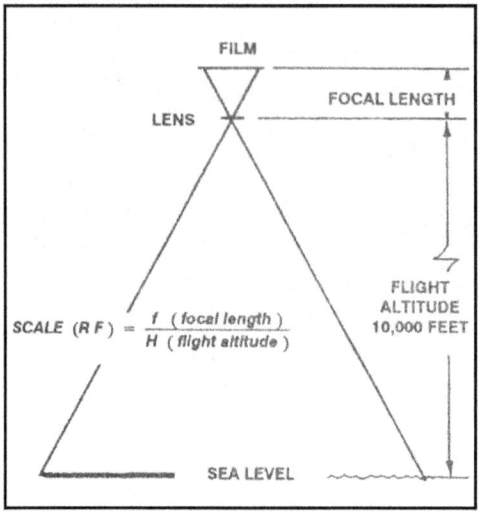

Figure 8-10. Basic computation of scale from sea level.

8-6. INDEXING

When aerial photos are taken of an area, it is convenient to have a record of the extent of coverage of each photo. A map on which the area covered by each photo is outlined and numbered or indexed to correspond to the photo is called an index map. There are two methods of preparing index maps.

  a. The four-corner method (Figures 8-11 and 8-12) requires location on the map of the exact point corresponding to each corner of the photo. If a recognizable object, such as a house or road junction can be found exactly at one of the corners, this point may be used on the map as the corner of the photo. If recognizable objects cannot be found at the corners, then the edges of the photo should be outlined on the map by lining up two or more identifiable objects along each edge; the points where the edges intersect should be the exact corners of the photo. If the photo is not a perfect vertical, the area outlined on the map will not be a perfect square or rectangle. After the four sides are drawn on the map, the number of the photograph is written in the enclosed area for identification. This number should be placed in the same corner as it is on the photo.

FM 3-25.26

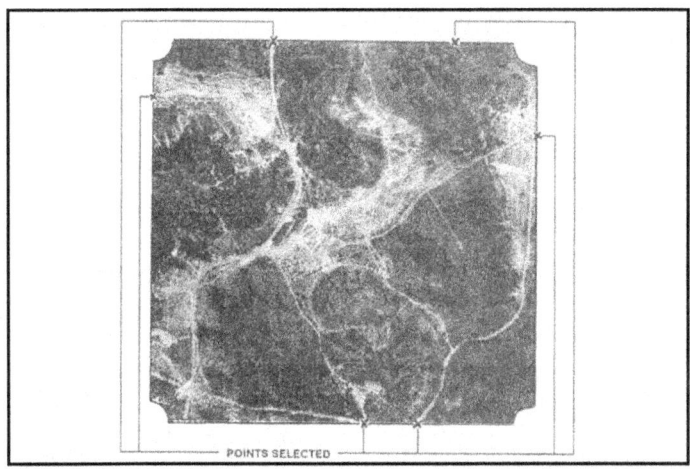

Figure 8-11. Four-corner method (selection of points).

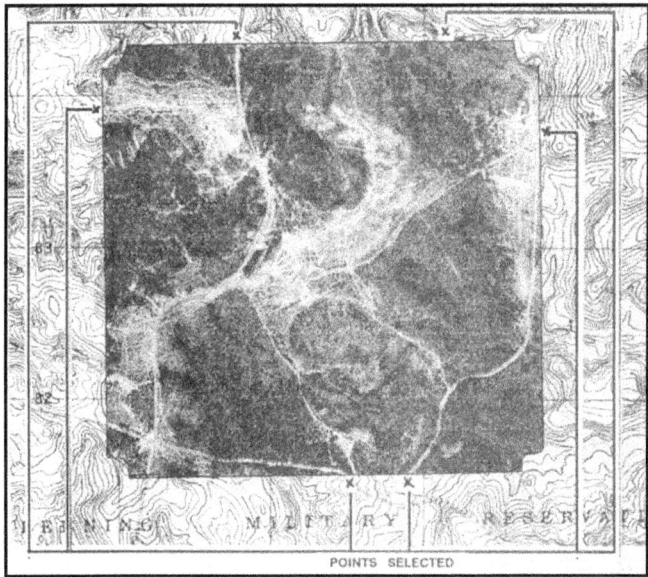

Figure 8-12. Plotting, using the four-corner method.

b. The template method is used when a large number of photos are to be indexed, and the exact area covered by each is not as important as approximate area and location. In this case, a template (cardboard pattern or guide) is cut to fit the average area the photos cover on the index map. It is used to outline the individual area covered by each photo.

(1) To construct a template, find the average map dimensions covered by the photos to be indexed as follows. Multiply the average length of the photos by the denominator of the average scale of the photos; multiply this by the scale of the map. Do the same for the width of the photos. This gives the average length and width of the area each photo covers on the map—or the size to which the template should be cut (Figure 8-13).

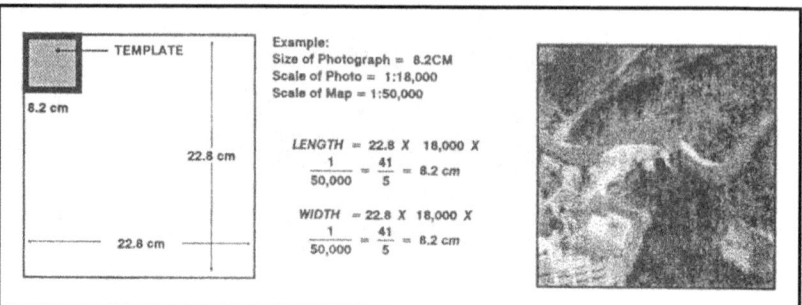

Figure 8-13. Constructing a template.

(2) To index the map, select the general area covered by the first photo and orient the photo to the map. Place the template over the area on the map and adjust it until it covers the area as completely and accurately as possible. Draw lines around the edges of the template. Remove the rectangle and proceed to the next photo (Figure 8-14).

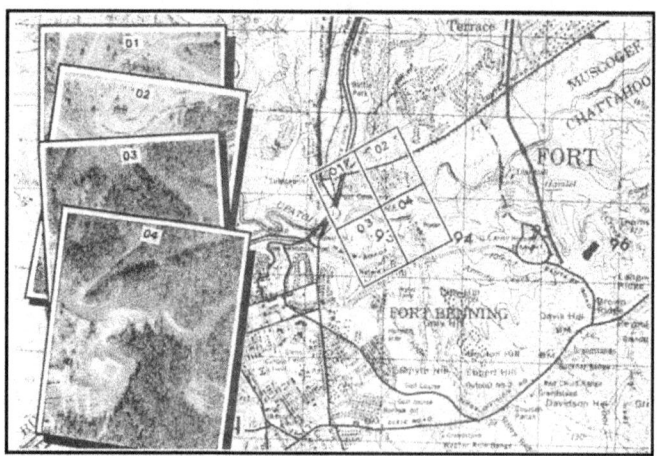

Figure 8-14. Indexing with a template.

c. After all photos have been plotted, write on the map sufficient information to identify the mission or sortie. If more than one sortie is plotted on one map or overlay, use a different color for each sortie.

d. In most cases, when a unit orders aerial photography, an index is included to give the basic information. Instead of being annotated on a map of the area, it appears on an overlay and is keyed to a map.

8-7. ORIENTATION OF PHOTOGRAPH

Orienting the photograph is important because it is of very little value as a map supplement or substitute if its location and direction are not known by the user.

  a. If a map of the same area as the photograph is available, the photograph is oriented to the map by comparing features common to both and then transferring a direction line from the map to the photograph.

  b. If no map is available, the shadows on a photograph may be used to get an approximate true-north line. This method is not recommended in the torrid zone (Figure 8-15).

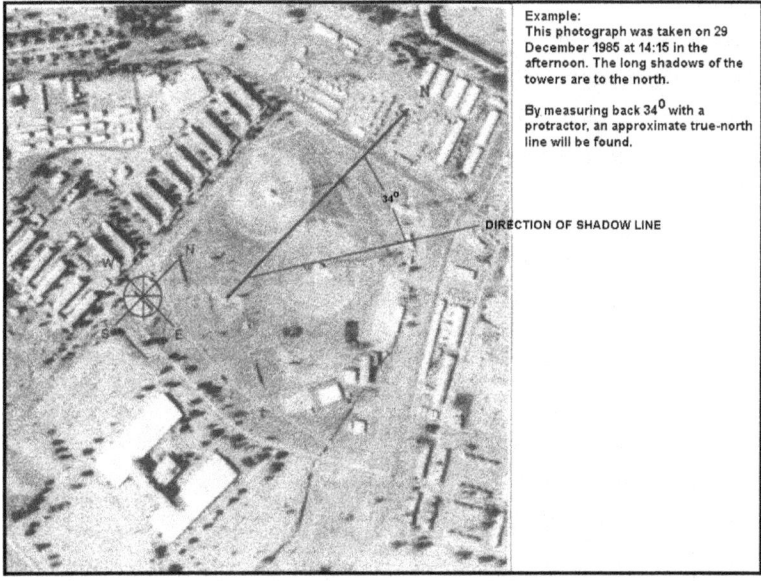

Figure 8-15. Using shadows on a photograph to find north.

  (1) *North Temperate Zone.* The sun moves from the east in the morning through south at noon to west in the afternoon. Conversely, shadow fall varies from west through north to east. Before noon, therefore, north is to the right of the direction of shadow fall; at noon, north is the direction of shadow fall; and after noon, north is to the left of shadow fall. On an average, the amount of variation in shadow fall per hour is 15 degrees. From marginal information, determine the number of hours from noon that the photo was taken and multiply that number by 15 degrees. With a protractor, measure an angle of that amount in the proper direction (right to left) from a clear, distinct shadow, and north is obtained. For photographs taken within three hours of noon, a reasonable accurate north direction can be obtained.

FM 3-25.26

Beyond these limits, the 15 degrees must be corrected, depending on time of year and latitude.

(2) *South Temperate Zone*. The sun moves from east through north at noon to west. Shadows then vary from west through south to east. Before noon, south is to the left of shadow fall; at noon, south is shadow fall; and after noon, south is to the right of shadow fall. Proceed as in (1) above to determine the direction of south.

c. On a photograph that can be oriented to the surrounding ground features by inspection, a magnetic-north line can be established using a compass.

(1) Orient the photograph by inspection.

(2) Open the compass and place it on the photograph.

(3) Without moving the photograph, rotate the compass until the north arrow is under the black fixed index line.

(4) Draw a line along the straight edge of the compass. This is a magnetic-north line.

8-8. POINT DESIGNATION GRID

Since aerial photographs are seldom exactly the same scale as a map of the same area, it is not feasible to print military grids on them. A special grid is used for the designation of points on photographs (Figure 8-16). This grid, known as the point designation grid, has no relation to the scale of the photo, to any direction, or to the grid used on any other photograph or map. It has only one purpose, to designate points on photographs.

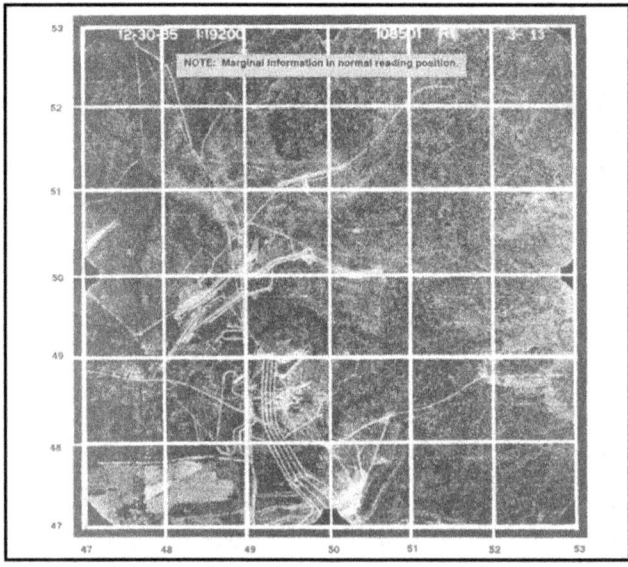

Figure 8-16. Point designation grid.

a. The point designation grid is rarely printed on photographs; therefore, it becomes the responsibility of each user to construct the grid on the photograph. All users must construct the grid in exactly the same way. Before the grid can be constructed or used, the photograph must be held so that the marginal information, regardless of where it is located, is in the normal reading position (Figure 8-17, step 1).

(1) Draw lines across the photograph joining opposite reference marks at the center of each photograph (fiducial marks). If there are no fiducial marks, the center of each side of the photograph is assumed to be the location of the marks (Figure 8-17, step 2).

(2) Space grid lines, starting with the center line, 4 centimeters (1.575 inches) apart (a distance equal to 1,000 meters at a scale of 1:25,000). The 1:25,000 map coordinate scale can be used for this dimension and to accurately designate points on the photograph, but this does not mean that distance can be scaled from the photograph. Extend the grid past the margins of the photograph so that a horizontal and vertical grid line fall outside the picture area (Figure 8-17, step 3).

(3) Number each center line "50" and give numerical values to the remaining horizontal and vertical lines so that they increase to the right and up (Figure 8-17, step 4).

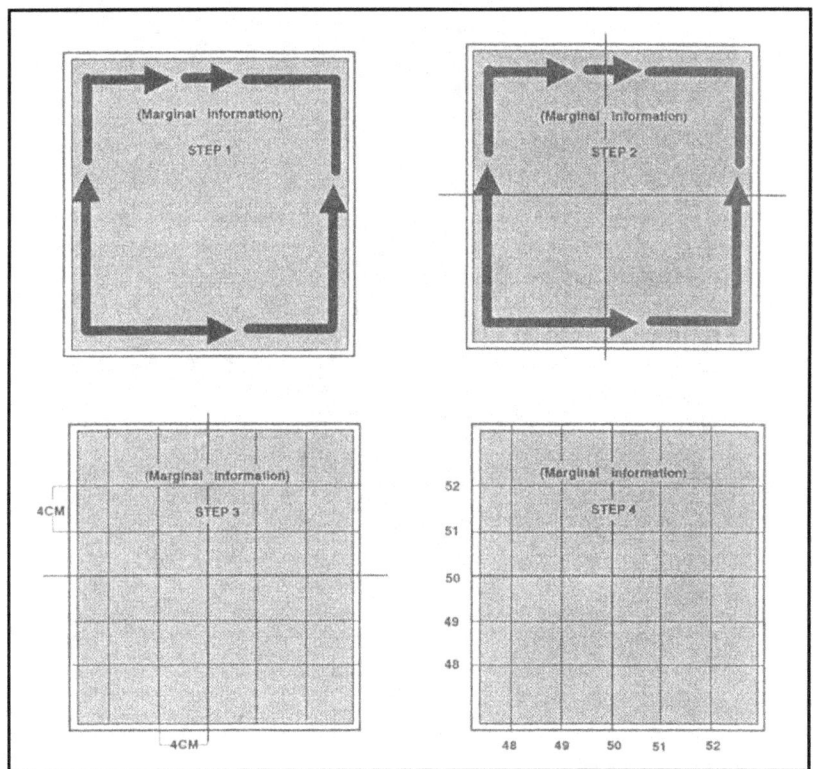

Figure 8-17. Constructing a point designation grid.

b. Once the photograph is oriented, the point designation grid is used in the same manner as the grid on a map (Figure 8-18), *read right and up*. The coordinate scale used with the UTM grid on maps at the scale of 1:25,000 may be used to subdivide the grid square in the same manner as on a map. However, because the same point designation grid is used on all photographs, the coordinates of a point on the photograph must be prefixed by the identifying marginal information of the photograph.

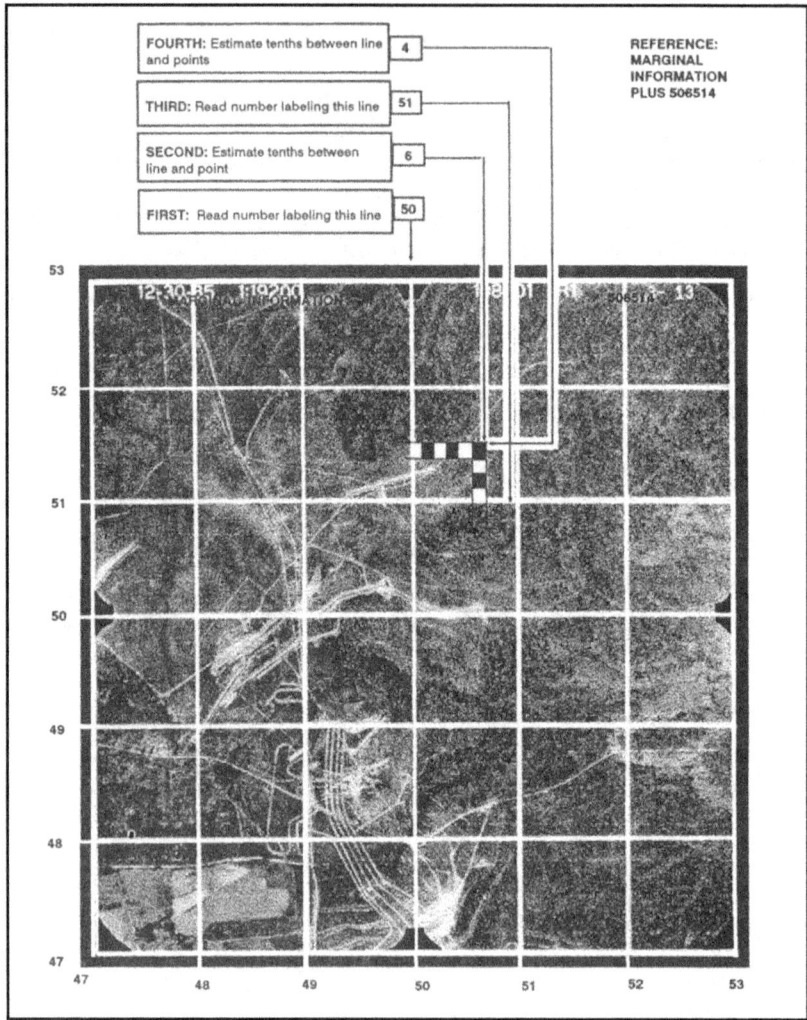

Figure 8-18. Reading point designation grid coordinates.

c. A grid coordinate using the point designation grid (Figure 8-19) consists of three parts:
- The letters "PDG" to indicate an aerial photograph rather than a map grid coordinate.

- The mission and photo negative number to identify which photograph is being used.
- The six numerical digits to locate the actual point on the photograph.

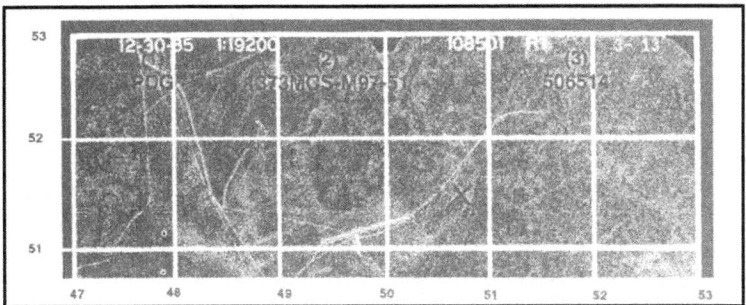

Figure 8-19. Locating the grid coordinate on a point designation grid.

8-9. IDENTIFICATION OF PHOTOGRAPH FEATURES

The identification of features on a photograph is not difficult if the following facts are remembered. The view that is presented by the aerial photograph is from above and, as a result, objects do not look familiar. Objects that are greatly reduced in size appear distorted. Most aerial photography is black and white, and all colors appear on the photograph in shades of gray. Generally speaking, the darker the natural color, the darker it will appear on the photograph.

   a. The identification of features on aerial photographs depends upon a careful application of five factors of recognition. No one factor will give a positive identification; all five are required.

   (1) *Size*. The size of unknown objects on a photograph, as determined from the scale of the photograph or a comparison with known objects of known size, gives a clue to their identity. For example, in a built-up area the smaller buildings are usually dwellings, and the larger buildings are commercial or community buildings.

   (2) *Shape (Pattern)*. Many features possess characteristic shapes that readily identify the features. Man-made features appear as straight or smooth curved lines, while natural features usually appear to be irregular. Some of the most prominent man-made features are highways, railroads, bridges, canals, and buildings. Compare the regular shapes of these to the irregular shapes of such natural features as streams and timber lines.

   (3) *Shadows*. Shadows are very helpful in identifying features since they show the familiar side view of the object. Some excellent examples are the shadows of water towers or smoke stacks. As viewed directly from above, only a round circle or dot is seen, whereas the shadow shows the profile and helps to identify the object. Relative lengths of shadows also usually give a good indication of relative heights of objects.

   (4) *Shade (Tone or Texture)*. Of the many different types of photographic film in use today, the film used for most aerial photography, except for special purposes, is panchromatic film. Panchromatic film is sensitive to all the colors of the spectrum; it registers them as shades of gray, ranging from white to black. This lighter or darker shade of features on aerial photographs is known as the tone. The tone is also dependent on the

texture of the features; a paved highway has a smooth texture and produces an even tone on the photograph, while a recently plowed field or a marsh has a rough, choppy texture and results in a rough or grainy tone. It is also important to remember that similar features may have different tones on different photographs, depending on the reflection of sunlight. For example, a river or body of water appears light if it is reflecting sunlight directly toward the camera, but appears dark otherwise. Its texture may be smooth or rough, depending on the surface of the water itself. As long as the variables are kept in mind, tone and texture may be used to great advantage.

(5) *Surrounding Objects*. Quite often an object not easily recognized by itself may be identified by its relative position to surrounding objects. Large buildings located beside railroads or railroad sidings are usually factories or warehouses. Identify schools by the baseball or football fields. It would be hard to tell the difference between a water tower next to a railroad station and a silo next to a barn unless the surrounding objects, such as the railroad tracks or cultivated fields, were considered.

b. Before a vertical photograph can be studied or used for identification of features, it must be oriented. This orienting is different from the orienting required for the construction or use of the point designation grid. Orienting for study consists of rotating the photograph so that the shadows on the photograph point toward yourself. You then face a source of light. This places the source of light, an object, and its shadow in a natural relationship. Failure to orient a photograph properly may cause the height or depth of an object to appear reversed. For example, a mine or quarry may appear to be a hill instead of a depression.

8-10. STEREOVISION

One of the limitations of the vertical aerial photograph is the lack of apparent relief. Stereoscopic vision (or as it is more commonly known, stereovision or depth perception) is the ability to see three-dimensionally or to see length, width, and depth (distance) at the same time. This requires two views of a single object from two slightly different positions. Most people have the ability to see three-dimensionally. Whenever an object is viewed, it is seen twice—once with the left eye and once with the right eye. The fusion or blending together of these two images in the brain permits the judgment of depth or distance.

a. In taking aerial photographs, it is rare for only a single picture to be taken. Generally, the aircraft flies over the area to be photographed taking a series of pictures, each of which overlaps the photograph preceding it and the photograph following it so that an unbroken coverage of the area is obtained (Figure 8-20). The amount of overlap is usually 56 percent, which means that 56 percent of the ground detail appearing on one photo also appears on the next photograph. When a single flight does not give the necessary coverage of an area, additional flights must be made. These additional flights are parallel to the first and must have an overlap between them. This overlap between flights is known as side lap and usually is between 15 and 20 percent (Figure 8-21).

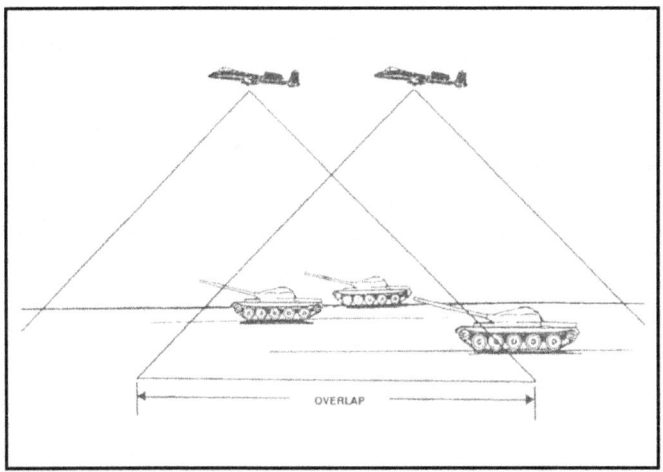

Figure 8-20. Photographic overlap.

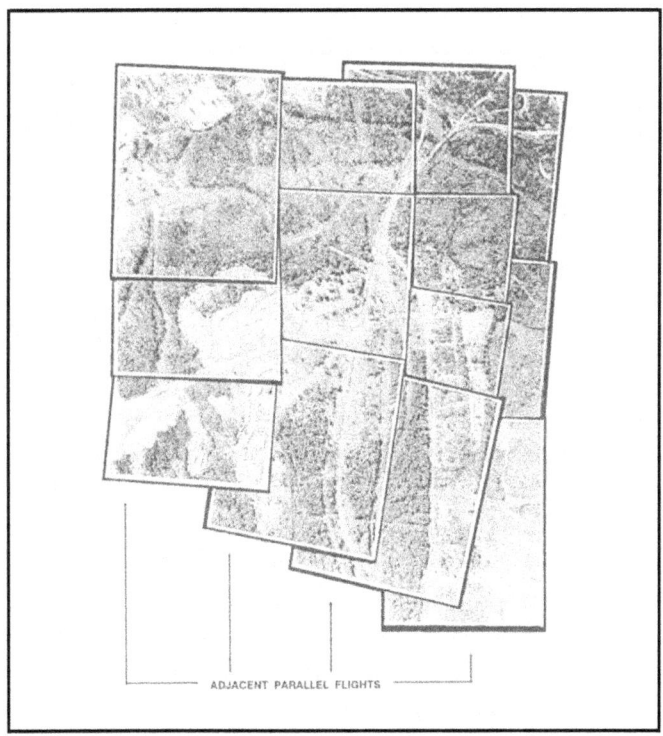

Figure 8-21. Side lap.

b. The requirement for stereovision can be satisfied by overlapping photographs if one eye sees the object on one photograph and the other eye sees the same object on another photograph. While this can be done after practice with the eyes alone, it is much easier if an optical aid is used. These optical aids are known as stereoscopes. There are many types of stereoscopes, but only the two most commonly used are discussed in this manual.

(1) *Pocket Stereoscope.* The pocket stereoscope (Figure 8-22), sometimes known as a lens stereoscope, consists of two magnifying lenses mounted in a metal frame. Because of its simplicity and ease of carrying, it is the type used most frequently by military personnel.

Figure 8-22. Pocket stereoscope.

(2) *Mirror Stereoscope.* The mirror stereoscope (Figure 8-23) is larger, heavier, and more subject to damage than the pocket stereoscope. It consists of four mirrors mounted in a metal frame.

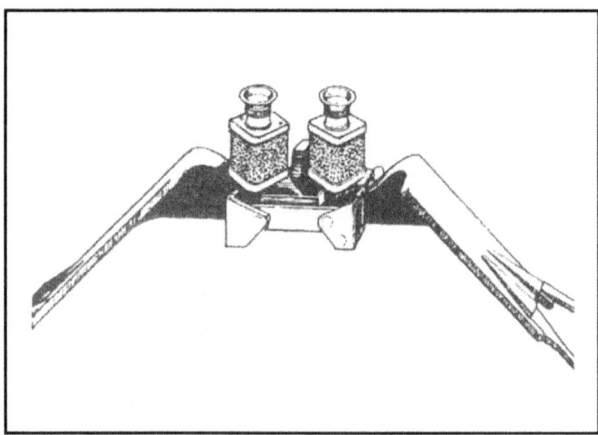

Figure 8-23. Mirror stereoscope.

c. One method to orient a pair of aerial photographs for the best three-dimensional viewing is as follows (Figure 8-24):

(1) Arrange the selected pair of photos in such a way that the shadows on them generally appear to fall toward the viewer. It is also desirable that the light source enters the side away from the observer during the study of the photographs.

(2) Place the pair of photographs on a flat surface so that the detail on one photograph is directly over the same detail on the other photograph.

(3) Place the stereoscope over the photographs so that the left lens is over the left photograph and the right lens is over the right photograph.

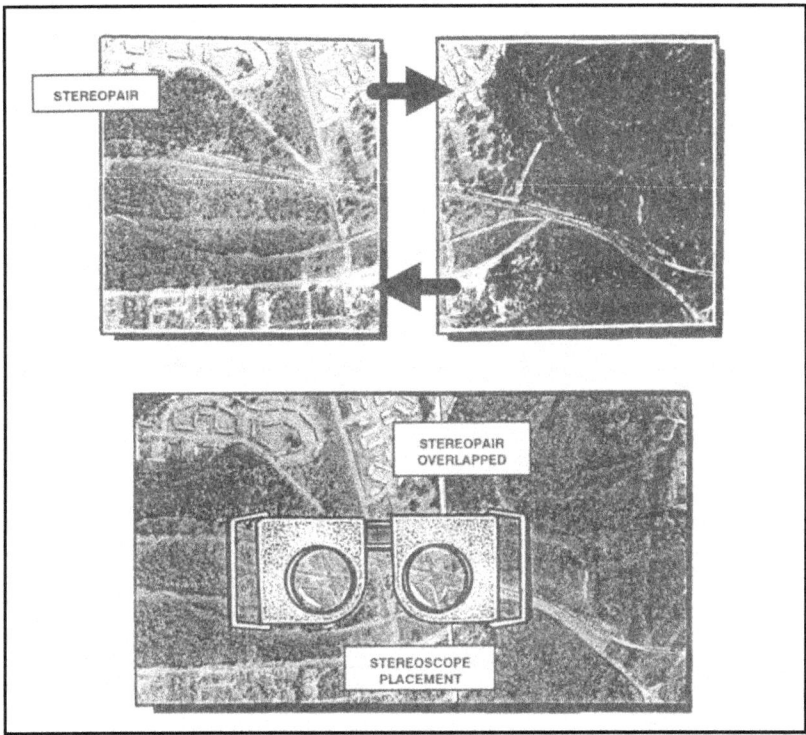

Figure 8-24. Placement of stereoscope over stereopair.

(4) Separate the photographs along the line of flight until a piece of detail appearing in the overlap area of the left photograph is directly under the left lens and the same piece of detail on the right photo is directly under the right lens.

(5) With the photograph and stereoscope in this position, a three-dimensional image should be seen. A few minor adjustments may be necessary such as adjusting the aerial photographs of the stereoscope to obtain the correct position for your eyes. The hills appear to rise and the valleys sink so that there is the impression of being in an aircraft looking down at the ground.

(6) The identification of features on photographs is much easier and more accurate with this three-dimensional view. The same five factors of recognition (size, shape, shadow, tone,

and surrounding objects) must still be applied; but now, with the addition of relief, a more natural view is seen.

FM 3-25.26

PART TWO
LAND NAVIGATION

CHAPTER 9
NAVIGATION EQUIPMENT AND METHODS

Compasses are the primary navigation tools to use when moving in an outdoor world where there is no other way to find directions. Soldiers should be thoroughly familiar with the compass and its uses. Part One of this manual discussed the techniques of map reading. To complement these techniques, a mastery of field movement techniques is essential. This chapter describes the lensatic compass and its uses, and some of the field-expedient methods used to find directions when compasses are not available.

9-1. TYPES OF COMPASSES

The *lensatic compass* is the most common and simplest instrument for measuring direction. It is discussed in detail in paragraph 9-2. The *artillery M2 compass* is a special-purpose instrument designed for accuracy; it will be discussed in Appendix J. The *wrist/pocket compass* is a small magnetic compass that can be attached to a wristwatch band. It contains a north-seeking arrow and a dial in degrees. A *protractor* can be used to determine azimuths when a compass is not available. However, it should be noted that when using the protractor on a map, only grid azimuths are obtained.

9-2. LENSATIC COMPASS

The lensatic compass (Figure 9-1) consists of three major parts: the cover, the base, and the lens.

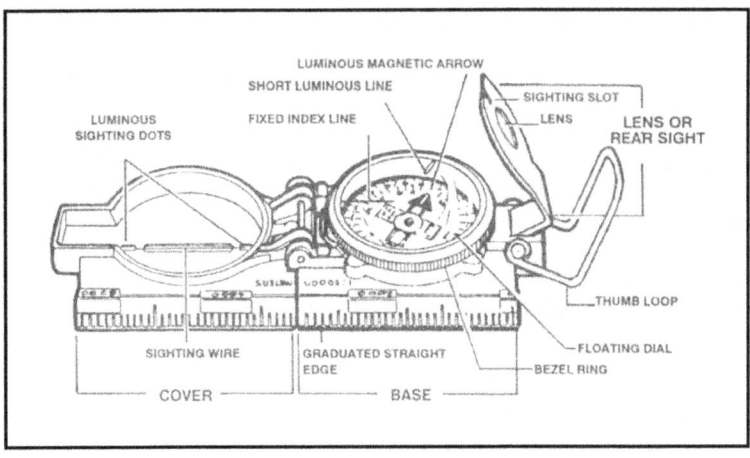

Figure 9-1. Lensatic compass.

FOUO
18 January 2005
9-1

a. **Cover.** The compass cover protects the floating dial. The cover contains the sighting wire (front sight) and two luminous sighting slots or dots used for night navigation.

b. **Base.** The base of the compass contains the following movable parts:

(1) The floating dial is mounted on a pivot so it can rotate freely when the compass is held level. Printed on the dial in luminous figures are an arrow and the letters E and W. The arrow always points to magnetic north and the letters fall at east (E) 90 degrees and west (W) 270 degrees on the dial. There are two scales; the outer scale denotes mils and the inner scale (normally in red) denotes degrees.

(2) Encasing the floating dial is a glass containing a fixed black index line.

(3) The bezel ring is a ratchet device that clicks when turned. It contains 120 clicks when rotated fully; each click is equal to 3 degrees. A short luminous line that is used in conjunction with the north-seeking arrow during navigation is contained in the glass face of the bezel ring.

(4) The thumb loop is attached to the base of the compass.

c. **Lens.** The lens is used to read the dial, and it contains the rear-sight slot used in conjunction with the front for sighting on objects. The rear sight also serves as a lock and clamps the dial when closed for its protection. The rear sight must be opened more than 45 degrees to allow the dial to float freely.

NOTE: When opened, the straight edge on the left side of the compass has a coordinate scale; the scale is 1:50,000 in newer compasses.

> **WARNING**
> Some older compasses will have a 1:25,000 scale. This scale can be used with a 1:50,000-scale map, but the values read must be halved. Check the scale.

9-3. COMPASS HANDLING

Compasses are delicate instruments and should be cared for accordingly.

a. **Inspection.** A detailed inspection is required when first obtaining and using a compass. One of the most important parts to check is the floating dial, which contains the magnetic needle. The user must also make sure the sighting wire is straight, the glass and crystal parts are not broken, the numbers on the dial are readable, and most important, that the dial does not stick.

b. **Effects of Metal and Electricity.** Metal objects and electrical sources can affect the performance of a compass. However, nonmagnetic metals and alloys do not affect compass readings. The following separation distances are suggested to ensure proper functioning of a compass:

High-tension power lines .... 55 meters.
Field gun, truck, or tank .... 18 meters.
Telegraph or telephone wires and barbed wire .... 10 meters.
Machine gun .... 2 meters.

Steel helmet or rifle... 1/2 meter.

c. **Accuracy.** A compass in good working condition is very accurate. However, a compass has to be checked periodically on a known line of direction, such as a surveyed azimuth, using a declination station. Compasses with more than 3 degrees variation should not be used.

d. **Protection.** If traveling with the compass unfolded, make sure the rear sight is fully folded down onto the bezel ring. This will lock the floating dial and prevent vibration, as well as protect the crystal and rear sight from damage.

9-4. USING A COMPASS

Magnetic azimuths are determined using magnetic instruments such as lensatic and M2 compasses. Employ the following techniques when using the lensatic compass.

a. **Using the Centerhold Technique.** First, open the compass to its fullest so that the cover forms a straightedge with the base. Move the lens (rear sight) to the rearmost position, allowing the dial to float freely. Next, place your thumb through the thumb loop, form a steady base with your third and fourth fingers, and extend your index finger along the side of the compass. Place the thumb of the other hand between the lens (rear sight) and the bezel ring; extend the index finger along the remaining side of the compass, and the remaining fingers around the fingers of the other hand. Pull your elbows firmly into your sides; this will place the compass between your chin and your belt. To measure an azimuth, simply turn your entire body toward the object, pointing the compass cover directly at the object. Once you are pointing at the object, look down and read the azimuth from beneath the fixed black index line (Figure 9-2). This preferred method offers the following advantages over the sighting technique:

- It is faster and easier to use.
- It can be used under all conditions of visibility.
- It can be used when navigating over any type of terrain.
- It can be used without putting down the rifle; however, the rifle must be slung well back over either shoulder.
- It can be used without removing eyeglasses.

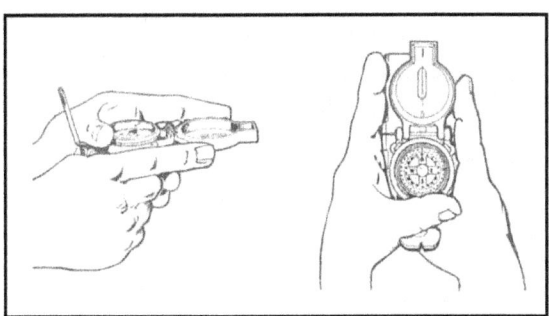

Figure 9-2. Centerhold technique.

b. **Using the Compass-to-Cheek Technique.** Fold the cover of the compass containing the sighting wire to a vertical position; then fold the rear sight slightly forward. Look through the rear-sight slot and align the front-sight hairline with the desired object in the distance. Glance down at the dial through the eye lens to read the azimuth (Figure 9-3).

NOTE: The compass-to-cheek technique is used almost exclusively for sighting, and it is the best technique for this purpose.

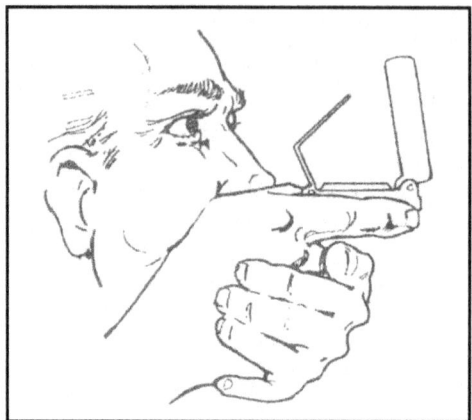

Figure 9-3. Compass-to-cheek technique.

   c. **Presetting a Compass and Following an Azimuth.** Although different models of the lensatic compass vary somewhat in the details of their use, the principles are the same.
   (1) During daylight hours or with a light source—
   (a) Hold the compass level in the palm of the hand.
   (b) Rotate it until the desired azimuth falls under the fixed black index line (for example, 320 degrees), maintaining the azimuth as prescribed (Figure 9-4).
   (c) Turn the bezel ring until the luminous line is aligned with the north-seeking arrow. Once the alignment is obtained, the compass is preset.
   (d) To follow an azimuth, assume the centerhold technique and turn your body until the north-seeking arrow is aligned with the luminous line. Proceed forward in the direction of the front cover's sighting wire, which is aligned with the fixed black index line that contains the desired azimuth.

FM 3-25.26

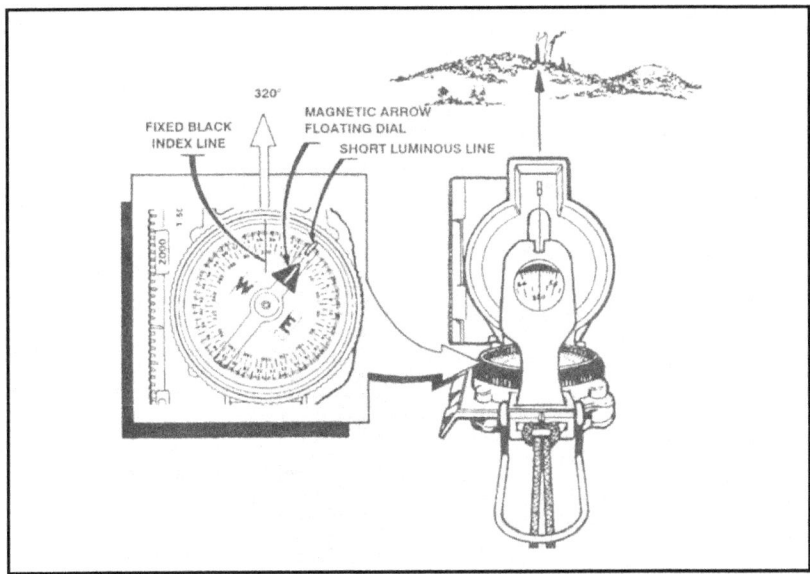

Figure 9-4. Compass preset at 320 degrees.

(2) During limited visibility, an azimuth may be set on the compass by the click method. Remember that the bezel ring contains 3-degree intervals (clicks).

(a) Rotate the bezel ring until the luminous line is over the fixed black index line.

(b) Find the desired azimuth and divide it by three. The result is the number of clicks that you have to rotate the bezel ring.

(c) Count the desired number of clicks. If the desired azimuth is smaller than 180 degrees, the number of clicks on the bezel ring should be counted in a counterclockwise direction. For example, the desired azimuth is 51 degrees; 51 degrees ÷ 3 = 17 clicks counterclockwise. If the desired azimuth is larger than 180 degrees, subtract the number of degrees from 360 degrees and divide by 3 to obtain the number of clicks. Count them in a clockwise direction. For example, the desired azimuth is 330 degrees; 360 degrees − 330 degrees = 30 ÷ 3 = 10 clicks clockwise.

(d) With the compass preset as described above, assume a centerhold technique and rotate your body until the north-seeking arrow is aligned with the luminous line on the bezel. Proceed forward in the direction of the front cover's luminous dots, which are aligned with the fixed black index line containing the azimuth.

(e) When the compass is to be used in darkness, an initial azimuth should be set while light is still available, if possible. With the initial azimuth as a base, any other azimuth that is a multiple of three can be established using the clicking feature of the bezel ring.

NOTE: Sometimes the desired azimuth is not exactly divisible by three, causing an option of rounding up or rounding down. Rounding up causes an increase in the value of the azimuth, and the object is to be found on the left. Rounding down

FOUO 9-5
18 January 2005

causes a decrease in the value of the azimuth, and the object is to be found on the right.

d. **Bypassing an Obstacle.** To bypass enemy positions or obstacles and still stay oriented, detour around the obstacle by moving at right angles for specified distances.

(1) For example, while moving on an azimuth of 90 degrees change your azimuth to 180 degrees and travel for 100 meters. Change your azimuth to 90 degrees and travel for 150 meters. Change your azimuth to 360 degrees and travel for 100 meters. Then, change your azimuth to 90 degrees and you are back on your original azimuth line (Figure 9-5).

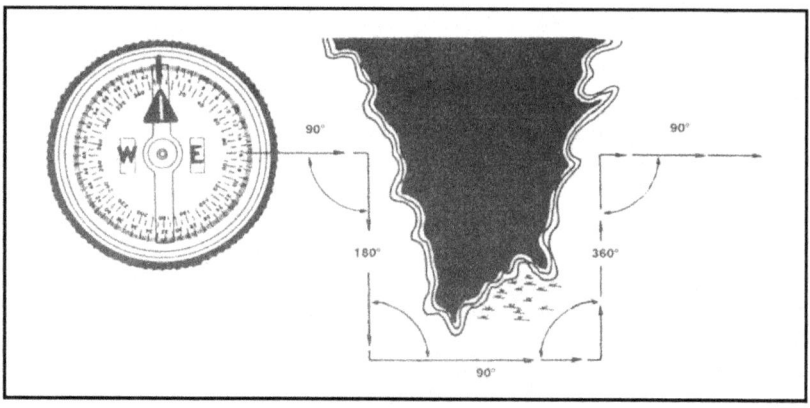

Figure 9-5. Bypassing an obstacle.

(2) Bypassing an unexpected obstacle at night is a fairly simple matter. To make a 90-degree turn to the right, hold the compass in the centerhold technique; turn until the center of the luminous letter E is under the luminous line (*do not* move the bezel ring). To make a 90-degree turn to the left, turn until the center of the luminous letter W is under the luminous line. This does not require changing the compass setting (bezel ring), and it ensures accurate 90-degree turns.

e. **Offset.** A deliberate offset is a planned magnetic deviation to the right or left of an azimuth to an objective. Use it when the objective is located along or in the vicinity of a linear feature such as a road or stream. Because of errors in the compass or in map reading, the linear feature may be reached without knowing whether the objective lies to the right or left. A deliberate offset by a known number of degrees in a known direction compensates for possible errors and ensures that upon reaching the linear feature, the user knows whether to go right or left to reach the objective. Ten degrees is an adequate offset for most tactical uses. Each degree offset moves the course about 18 meters to the right or left for each 1,000 meters traveled. For example, in Figure 9-6, the number of degrees offset is 10. If the distance traveled to "X" in 1,000 meters, then "X" is located about 180 meters to the right of the objective.

FM 3-25.26

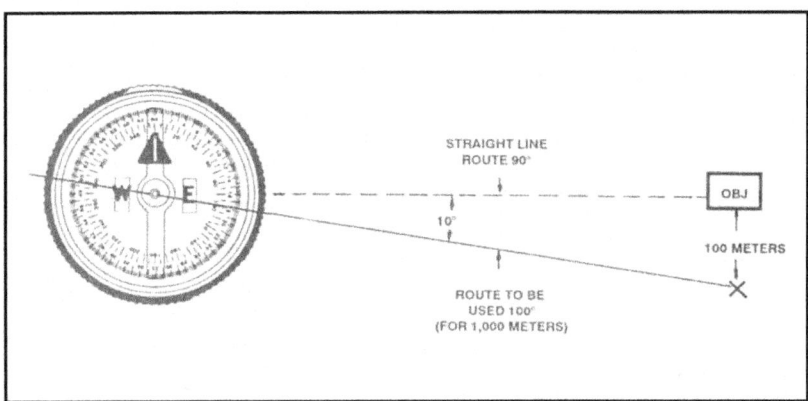

Figure 9-6. Deliberate offset to the objective.

9-5. FIELD-EXPEDIENT METHODS

When a compass is not available, different techniques may be used to determine the four cardinal directions.

    a. **Shadow-Tip Method.** This simple and accurate method of finding direction by the sun consists of four basic steps (Figure 9-7).

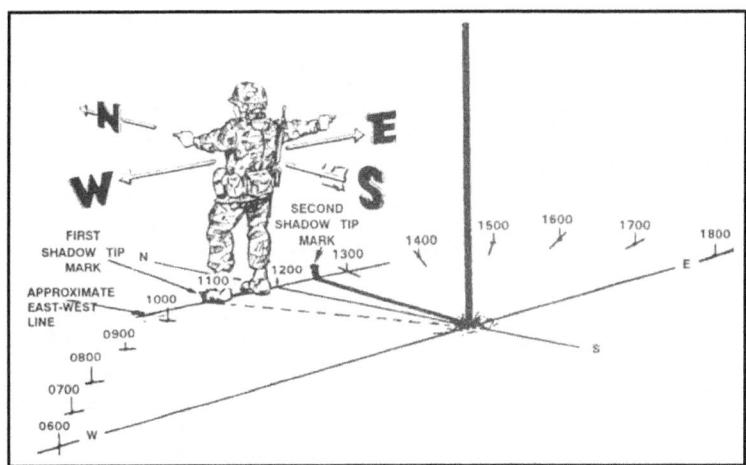

Figure 9-7. Determining directions and time by shadow.

Step 1.   Place a stick or branch into the ground at a level spot where a distinctive shadow will be cast. Mark the shadow tip with a stone, twig, or other means. This first shadow mark is always the west direction.

Step 2.   Wait 10 to 15 minutes until the shadow tip moves a few inches. Mark the new position of the shadow tip in the same way as the first.

Step 3. Draw a straight line through the two marks to obtain an approximate east-west line.

Step 4. Standing with the first mark (west) to your left, the other directions are simple; north is to the front, east is to the right, and south is behind you.

(1) A line drawn perpendicular to the east-west line at any point is the approximate north-south line. If you are uncertain which direction is east and which is west, observe this simple rule—the first shadow-tip mark is always in the west direction, everywhere on earth.

(2) The shadow-tip method can also be used as a shadow clock to find the approximate time of day (Figure 9-7).

(a) To find the time of day, move the stick to the intersection of the east-west line and the north-south line, and set it vertically in the ground. The west part of the east-west line indicates 0600 hours, and the east part is 1800 hours, anywhere on earth, because the basic rule always applies.

(b) The north-south line now becomes the noon line. The shadow of the stick is an hour hand in the shadow clock, and with it you can estimate the time using the noon line and the 6 o'clock line as your guides. Depending on your location and the season, the shadow may move either clockwise or counterclockwise, but this does not alter your manner of reading the shadow clock.

(c) The shadow clock is not a timepiece in the ordinary sense. It makes every day 12 unequal hours long, and always reads 0600 hours at sunrise and 1800 hours at sunset. The shadow clock time is closest to conventional clock time at midday, but the spacing of the other hours compared to conventional time varies somewhat with the locality and the date. However, it does provide a satisfactory means of telling time in the absence of properly set watches.

(d) The shadow-tip system is not intended for use in polar regions, which the Department of Defense defines as being above 60 degrees latitude in either hemisphere. Distressed persons in these areas are advised to stay in one place so that search/rescue teams can easily find them. The presence and location of all aircraft and ground parties in polar regions are reported to and checked regularly by governmental or other agencies, and any need for help becomes quickly known.

b. **Watch Method.** A watch can be used to determine the approximate true north and true south.

(1) In the north temperate zone only, the hour hand is pointed toward the sun. A south line can be found midway between the hour hand and 1200 hours, standard time. If on daylight savings time, the north-south line is found between the hour hand and 1300 hours. If there is any doubt as to which end of the line is north, remember that the sun is in the east before noon and in the west after noon.

(2) The watch may also be used to determine direction in the south temperate zone; however, the method is different. The 1200-hour dial is pointed toward the sun, and halfway between 1200 hours and the hour hand will be a north line. If on daylight savings time, the north line lies midway between the hour hand and 1300 hours (Figure 9-8).

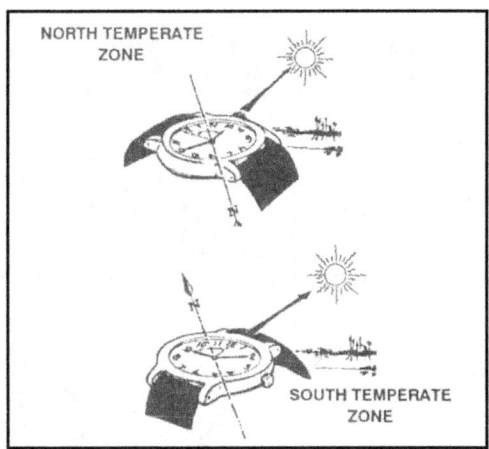

Figure 9-8. Determining direction using a watch.

(3) The watch method can be in error, especially in the lower latitudes, and may cause *circling*. To avoid this, make a shadow clock and set your watch to the time indicated. After traveling for an hour, take another shadow-clock reading. Reset your watch if necessary.

c. **Star Method.** Less than 60 of about 5,000 stars visible to the eye are used by navigators. The stars seen as we look up at the sky at night are not evenly scattered across the whole sky. Instead they are in groups called constellations.

(1) The constellations that we see depends partly on where we are located on the earth, the time of the year, and the time of the night. The night changes with the seasons because of the journey of the earth around the sun, and it also changes from hour to hour because the turning of the earth makes some constellations seem to travel in a circle. But there is one star that is in almost exactly the same place in the sky all night long every night. It is the North Star, also known as the Polar Star or Polaris.

(a) The North Star is less than 1 degree off true north and does not move from its place because the axis of the earth is pointed toward it. The North Star is in the group of stars called the Little Dipper. It is the last star in the handle of the dipper. There are two stars in the Big Dipper, which are a big help when trying to find the North Star. They are called the Pointers, and an imaginary line drawn through them five times their distance points to the North Star.

(b) Many stars are brighter than the North Star, but none is more important because of its location. However, the North Star can only be seen in the northern hemisphere so it cannot serve as a guide south of the equator. The farther one goes north, the higher the North Star is in the sky, and above latitude 70 degrees, it is too high in the sky to be useful (Figure 9-9, page 9-10).

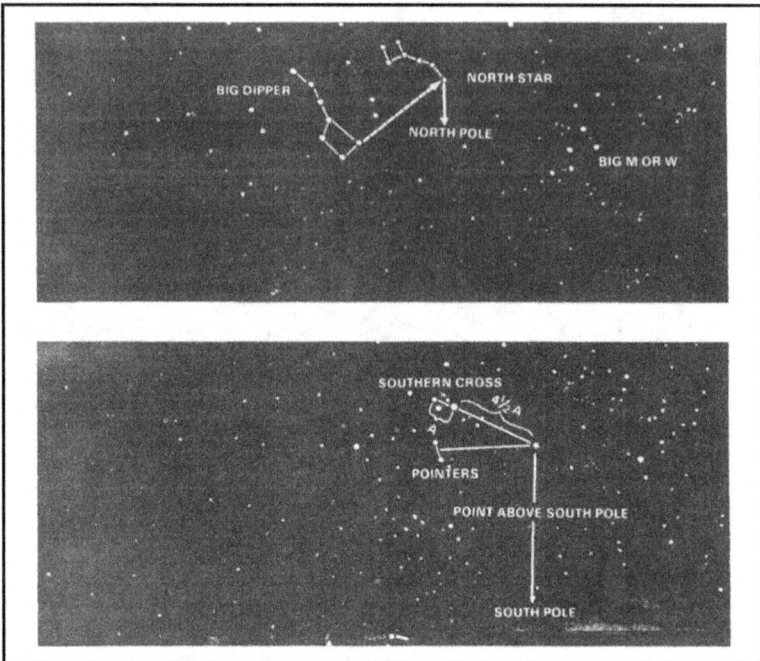

Figure 9-9. Determining direction by the North Star and Southern Cross.

(2) Depending on the star selected for navigation, azimuth checks are necessary. A star near the north horizon serves for about half an hour. When moving south, azimuth checks should be made every 15 minutes. When traveling east or west, the difficulty of staying on azimuth is caused more by the likelihood of the star climbing too high in the sky or losing itself behind the western horizon than it is by the star changing direction angle. When this happens, it is necessary to change to another guide star. The Southern Cross is the main constellation used as a guide south of the equator, and the general directions for using north and south stars are reversed. When navigating using the stars as guides, the user must know the different constellation shapes and their locations throughout the world (Figure 9-10 and Figure 9-11 on page 9-12).

FM 3-25.26

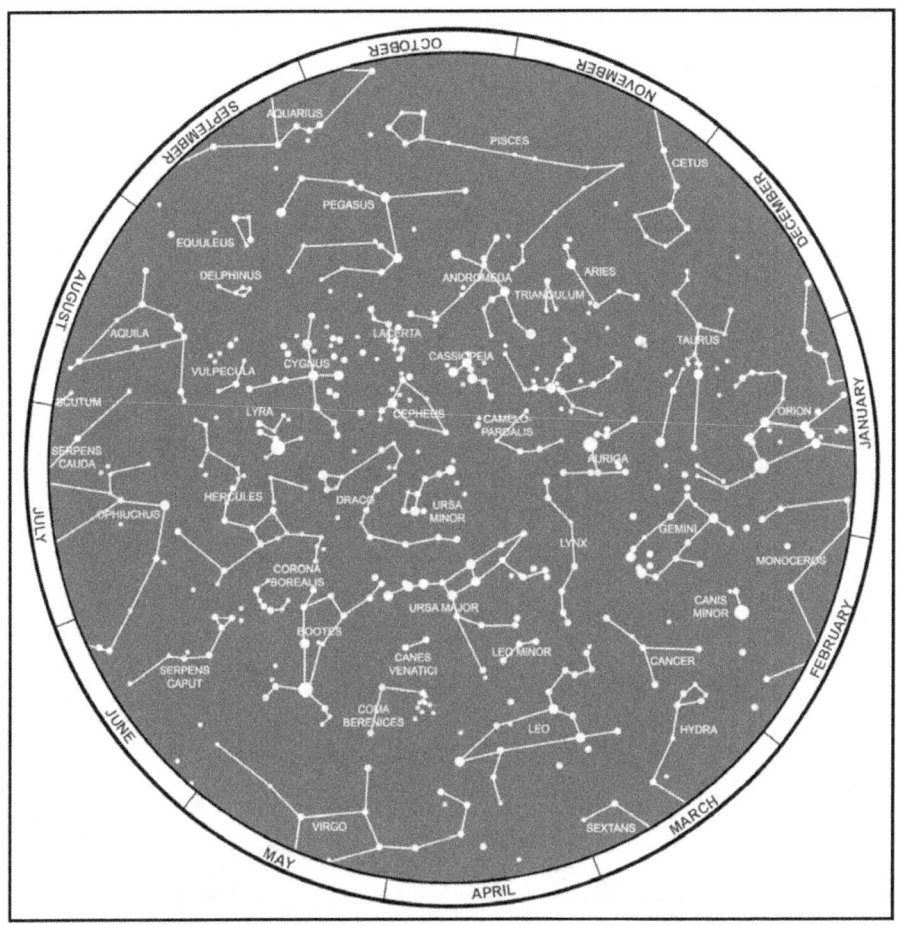

Figure 9-10. Constellations, northern hemisphere.

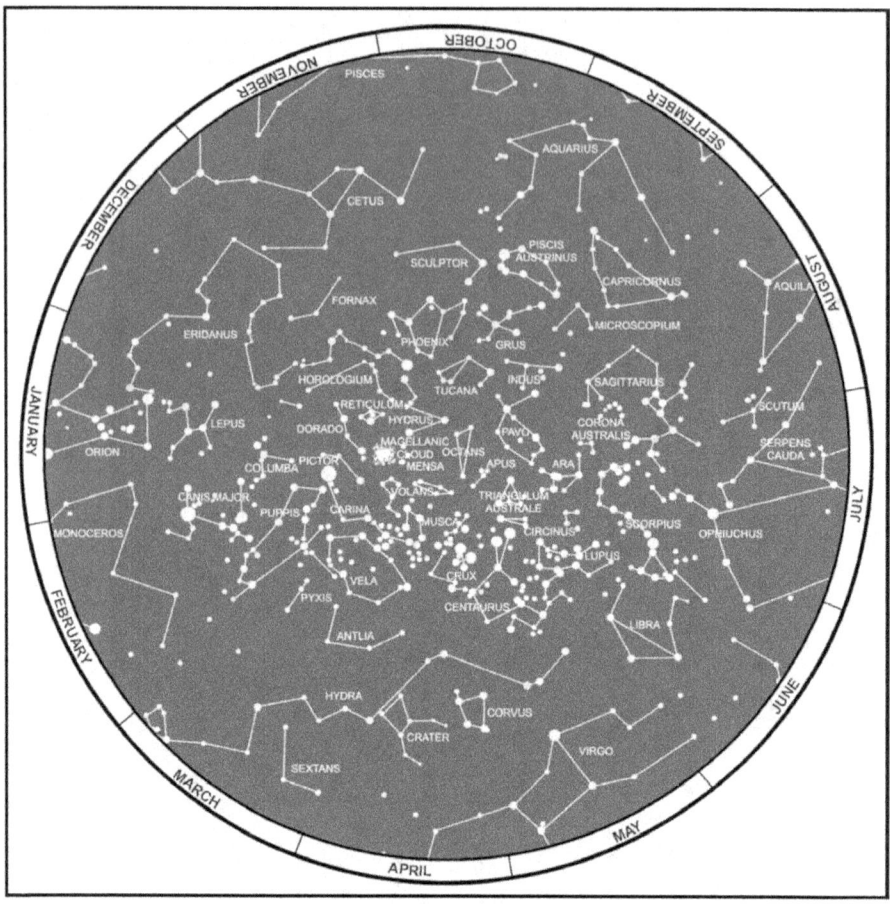

Figure 9-11. Constellations, southern hemisphere.

9-6. GLOBAL POSITIONING SYSTEM

The GPS is a space-based, global, all-weather, continuously available, radio positioning navigation system. It is highly accurate in determining position location derived from signal triangulation from a satellite constellation system. It is capable of determining latitude, longitude, and altitude of the individual user. It is being fielded in hand-held, manpack, vehicular, aircraft, and watercraft configurations. The GPS receives and processes data from satellites on either a simultaneous or sequential basis. It measures the velocity and range with respect to each satellite, processes the data in terms of an earth-centered, earth-fixed coordinate system, and displays the information to the user in geographic or military grid coordinates. (See Appendix I for more information on the GPS.)

    a. The GPS can provide precise steering information, as well as position location. The receiver can accept many checkpoints entered in any coordinate system by the user and convert them to the desired coordinate system. The user then calls up the desired checkpoint and the receiver will display direction and distance to the checkpoint. The GPS does not have inherent drift, an improvement over the Inertial Navigation System, and the receiver will automatically update its position. The receiver can also compute time to the next checkpoint.

b. Specific uses for the GPS are position location; navigation; weapon location; target and sensor location; coordination of firepower; scout and screening operations; combat resupply; location of obstacles, barriers, and gaps; and communication support. The GPS also has the potential to allow units to train their Soldiers and provide the following:

- Performance feedback.
- Knowledge of routes taken by the Soldier.
- Knowledge of errors committed by the Soldier.
- Comparison of planned versus executed routes.
- Safety and control of lost and injured Soldiers.

This Page intentionally left blank.

CHAPTER 10
ELEVATION AND RELIEF

The elevation of points on the ground and the relief of an area affect the movement, positioning, and, in some cases, effectiveness of military units. Soldiers must know how to determine locations of points on a map, measure distances and azimuths, and identify symbols on a map. They must also be able to determine the elevation and relief of areas on standard military maps. To do this, they must first understand how the mapmaker indicated the elevation and relief on the map.

10-1. DEFINITIONS
There must be a reference or start point to measure anything. The reference or start point for vertical measurement of elevation on a standard military map is the *datum plane* or *mean sea level*, the point halfway between high tide and low tide. *Elevation* of a point on the earth's surface is the vertical distance it is above or below mean sea level. *Relief* is the representation (as depicted by the mapmaker) of the shapes of hills, valleys, streams, or terrain features on the earth's surface.

10-2. METHODS OF DEPICTING RELIEF
Mapmakers use several methods to depict relief of the terrain.

   a. **Layer Tinting.** Layer tinting is a method of showing relief by color. A different color is used for each band of elevation. Each shade of color, or band, represents a definite elevation range. A legend is printed on the map margin to indicate the elevation range represented by each color. However, this method does not allow the map user to determine the exact elevation of a specific point—only the range.

   b. **Form Lines.** Form lines are not measured from any datum plane. Form lines have no standard elevation and give only a general idea of relief. Form lines are represented on a map as dashed lines and are never labeled with representative elevations.

   c. **Shaded Relief.** Relief shading indicates relief by a shadow effect achieved by tone and color that results in the darkening of one side of terrain features such as hills and ridges. The darker the shading, the steeper the slope. Shaded relief is sometimes used in conjunction with contour lines to emphasize these features.

   d. **Hachures.** Hachures are short, broken lines used to show relief. Hachures are sometimes used with contour lines. They do not represent exact elevations, but are mainly used to show large, rocky outcrop areas. Hachures are used extensively on small-scale maps to show mountain ranges, plateaus, and mountain peaks.

   e. **Contour Lines.** Contour lines are the most common method of showing relief and elevation on a standard topographic map. A contour line represents an imaginary line on the ground, above or below sea level. All points on the contour line are at the same elevation. The elevation represented by contour lines is the vertical distance above or below sea level. The three types of contour lines (Figure 10-1, page 10-2) used on a standard topographic map are index, intermediate, and supplementary.

(1) ***Index.*** Starting at zero elevation or mean sea level, every fifth contour line is a heavier line. These are known as index contour lines. Normally, each index contour line is numbered at some point. This number is the elevation of that line.

(2) ***Intermediate.*** The contour lines falling between the index contour lines are called intermediate contour lines. These lines are finer and do not have their elevations given. There are normally four intermediate contour lines between index contour lines.

(3) ***Supplementary.*** These contour lines resemble dashes. They show changes in elevation of at least one-half the contour interval. Supplementary lines are normally found where there is very little change in elevation such as on fairly level terrain.

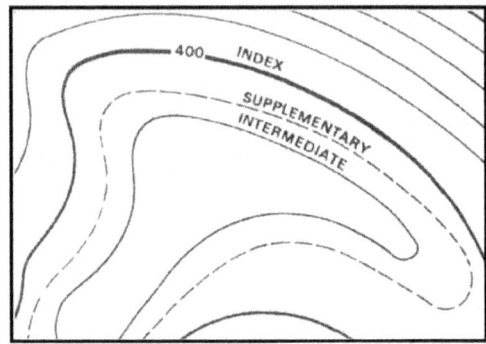

Figure 10-1. Contour lines.

10-3. CONTOUR INTERVALS

Before the elevation of any point on the map can be determined, the user must know the contour interval for the map he is using. The contour interval measurement given in the marginal information is the vertical distance between adjacent contour lines. Use the following procedures to determine the elevation of a point on the map.

    a. Determine the contour interval and the unit of measure used; for example, feet, meters, or yards (Figure 10-2).

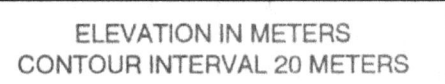

Figure 10-2. Example contour interval note.

b. Find the numbered index contour line nearest the point you are trying to determine the elevation for (Figure 10-3).

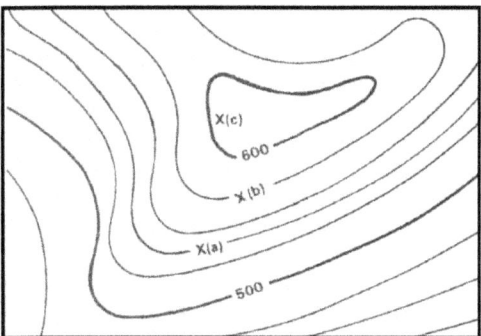

Figure 10-3. Points on contour lines.

c. Determine if you are going from lower elevation to higher, or vice versa. In Figure 10-3, point (a) is between the index contour lines. The lower index contour line is numbered 500, which means any point on that line is at an elevation of 500 meters above mean sea level. The upper index contour line is numbered 600, or 600 meters. Going from the lower to the upper index contour line shows an increase in elevation.

d. To determine the exact elevation of point (a), start at the index contour line numbered 500 and count the number of intermediate contour lines to point (a). Point (a) is located on the second intermediate contour line above the 500-meter index contour line. The contour interval is 20 meters (Figure 10-2), thus each intermediate contour line crossed to get to point (a) adds 20 meters to the 500-meter index contour line. The elevation of point (a) is 540 meters; the elevation has increased.

e. To determine the elevation of point (b), go to the nearest index contour line. In this case, it is the upper index contour line numbered 600. Point (b) is located on the intermediate contour line immediately below the 600-meter index contour line. Below means downhill or a lower elevation. Therefore, point (b) is located at an elevation of 580 meters. Remember, if you are increasing elevation, add the contour interval to the nearest index contour line. If you are decreasing elevation, subtract the contour interval from the nearest index contour line.

f. To determine the elevation to a hilltop, point (c), add one-half the contour interval to the elevation of the last contour line. In this example, the last contour line before the hilltop is an index contour line numbered 600. Add one-half the contour interval, 10 meters, to the index contour line. The elevation of the hilltop would be 610 meters.

g. There may be times when you need to determine the elevation of points to a greater accuracy. To do this, you must determine how far between the two contour lines the point lies. However, most military needs are satisfied by estimating the elevation of points between contour lines (Figure 10-4).

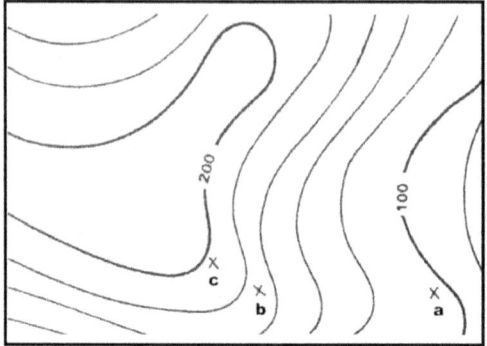

Figure 10-4. Points between contour lines.

(1) If the point is less than one-fourth the distance between contour lines, the elevation will be the same as the last contour line. In Figure 10-4, the elevation of *point a* will be 100 meters. To estimate the elevation of a point between one-fourth and three-fourths of the distance between contour lines, add one-half the contour interval to the last contour line.

(2) *Point b* is one-half the distance between contour lines. The contour line immediately below *point b* is at an elevation of 160 meters. The contour interval is 20 meters; thus one-half the contour interval is 10 meters. In this case, add 10 meters to the last contour line of 160 meters. The elevation of *point b* would be about 170 meters.

(3) A point located more than three-fourths of the distance between contour lines is considered to be at the same elevation as the next contour line. *Point c* is located three-fourths of the distance between contour lines. In Figure 10-4, *point c* would be considered to be at an elevation of 200 meters.

h. To estimate the elevation to the bottom of a depression, subtract one-half the contour interval from the value of the lowest contour line before the depression. In Figure 10-5, the lowest contour line before the depression is 240 meters in elevation. Thus, the elevation at the edge of the depression is 240 meters. To determine the elevation at the bottom of the depression, subtract one-half the contour interval. The contour interval for this example is 20 meters. Subtract 10 meters from the lowest contour line immediately before the depression. The result is that the elevation at the bottom of the depression is 230 meters. The tick marks on the contour line forming a depression always point to lower elevations.

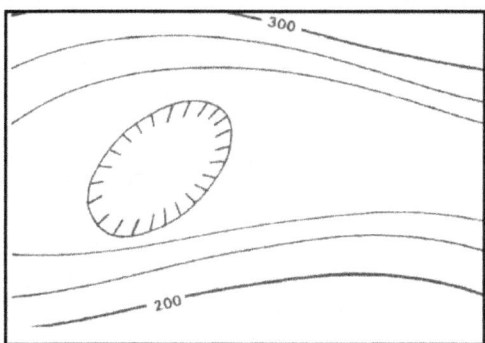

Figure 10-5. Depression.

i. In addition to the contour lines, bench marks and spot elevations are used to indicate points of known elevations on the map.

(1) Bench marks, the more accurate of the two, are symbolized by a black X; for example X BM 214. The 214 indicates that the center of the X is at an elevation of 214 units of measure (feet, meters, or yards) above mean sea level. To determine the units of measure, refer to the contour interval in the marginal information.

(2) Spot elevations are shown by a brown X and are usually located at road junctions and on hilltops and other prominent terrain features. If the elevation is shown in black numerals, it has been checked for accuracy; if it is in brown, it has not been checked.

NOTE: New maps are being printed using a dot instead of brown Xs.

10-4. TYPES OF SLOPES

The rate of rise or fall of a terrain feature is known as its slope. Depending on the military mission, Soldiers may need to determine not only the height of a hill, but the degree of the hill's slope as well. The speed at which equipment or personnel can move is affected by the slope of the ground or terrain feature. This slope can be determined from the map by studying the contour lines—the closer the contour lines, the steeper the slope; the farther apart the contour lines, the gentler the slope. Four types of slopes that concern the military are gentle, steep, concave, and convex.

a. **Gentle.** Contour lines showing a uniform, gentle slope will be evenly spaced and wide apart (Figure 10-6, page 10-6). Considering relief only, a uniform, gentle slope allows the defender to use grazing fire. The attacking force has to climb a slight incline.

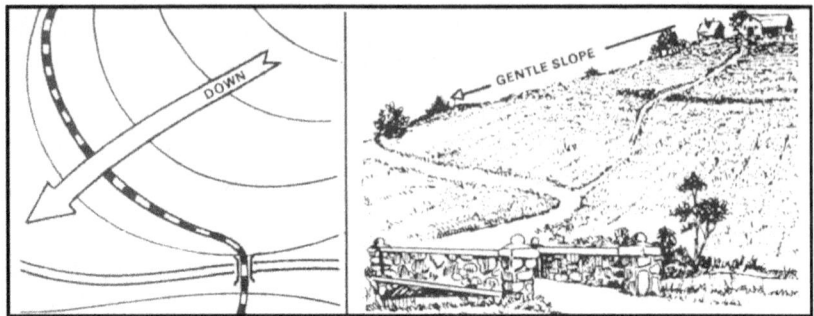

Figure 10-6. Uniform, gentle slope.

b. **Steep.** Contour lines showing a uniform, steep slope on a map will be evenly spaced, but close together. Remember, the closer the contour lines, the steeper the slope (Figure 10-7). Considering relief only, a uniform, steep slope allows the defender to use grazing fire, and the attacking force has to negotiate a steep incline.

Figure 10-7. Uniform, steep slope.

c. **Concave.** Contour lines showing a concave slope on a map will be closely spaced at the top of the terrain feature and widely spaced at the bottom (Figure 10-8). Considering relief only, the defender at the top of the slope can observe the entire slope and the terrain at the bottom, but he cannot use grazing fire. The attacker would have no cover from the defender's observation of fire, and his climb would become more difficult as he gets farther up the slope.

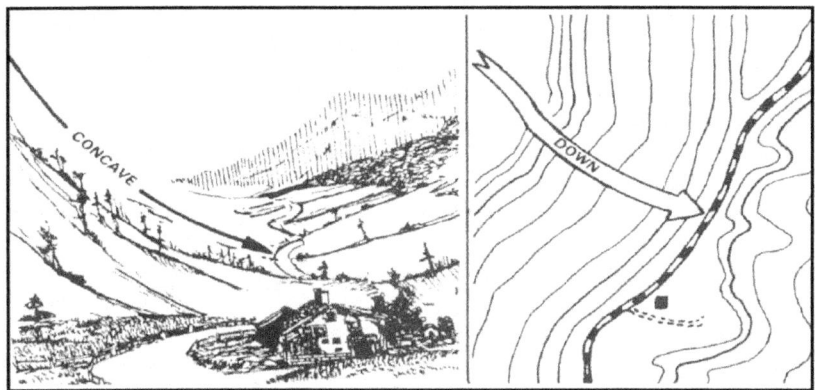

Figure 10-8. Concave slope.

d. **Convex.** Contour lines showing a convex slope on a map will be widely spaced at the top and closely spaced at the bottom (Figure 10-9). Considering relief only, the defender at the top of the convex slope can obtain a small distance of grazing fire, but he cannot observe most of the slope or the terrain at the bottom. The attacker will have concealment on most of the slope and an easier climb as he nears the top.

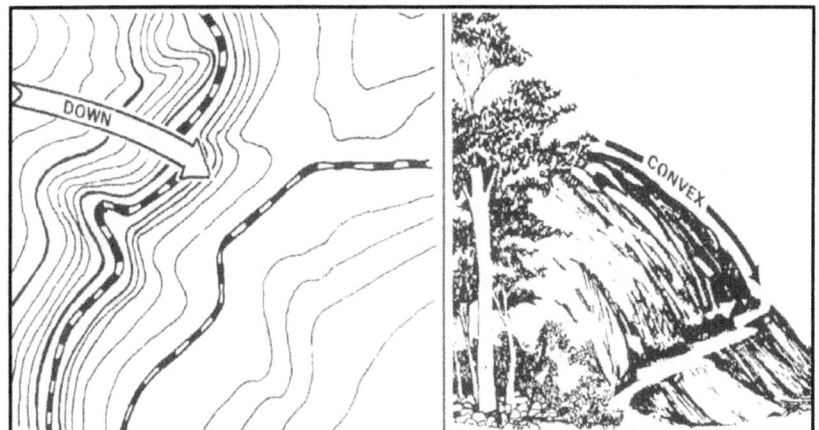

Figure 10-9. Convex slope.

10-5. PERCENTAGE OF SLOPE

The speed at which personnel and equipment can move up or down a hill is affected by the slope of the ground and the limitations of the equipment. Because of this, a more exact way of describing a slope is necessary.

a. Slope may be expressed in several ways, but all depend upon the comparison of vertical distance (VD) to horizontal distance (HD) (Figure 10-10, page 10-8). Before we can determine the percentage of a slope, we must know the VD of the slope. The VD is determined by subtracting the lowest point of the slope from the highest point. Use the contour lines to determine the highest and lowest point of the slope (Figure 10-11).

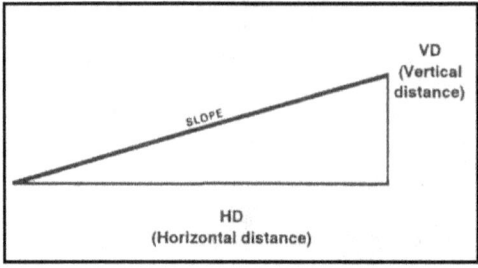

Figure 10-10. Slope diagram.

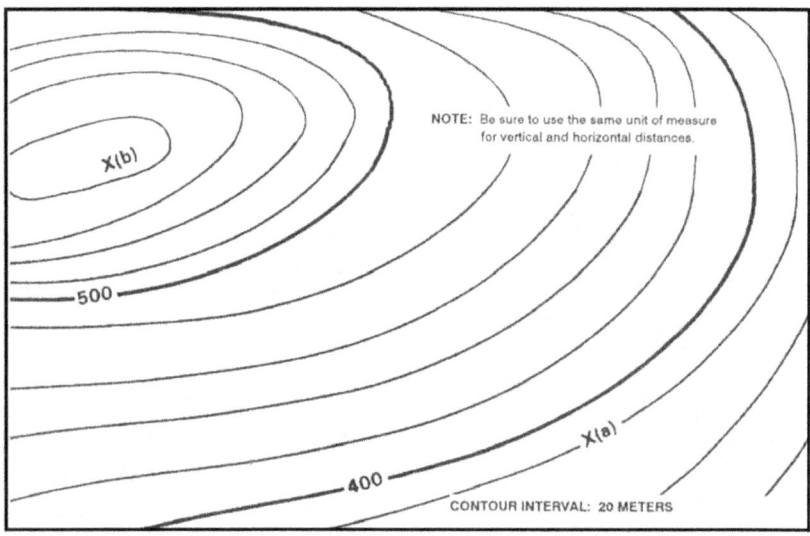

Figure 10-11. Contour lines around a slope.

b. To determine the percentage of the slope between points (a) and (b) in Figure 10-11, determine the elevation of point (b) (590 meters). Then determine the elevation of point (a) (380 meters). Determine the vertical distance between the two points by subtracting the elevation of point (a) from the elevation of point (b). The difference (210 meters) is the VD between points (a) and (b). Then measure the HD between the two points on the map in Figure 10-12. After the horizontal distance has been determined, compute the percentage of the slope by using the formula shown in Figure 10-13.

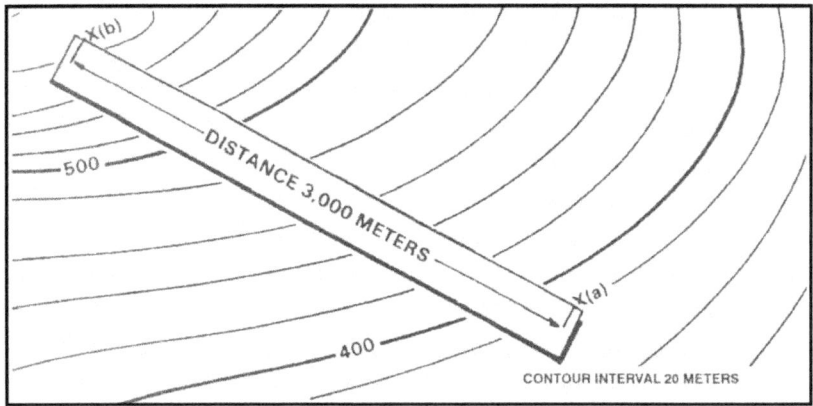

Figure 10-12. Measuring horizontal distance.

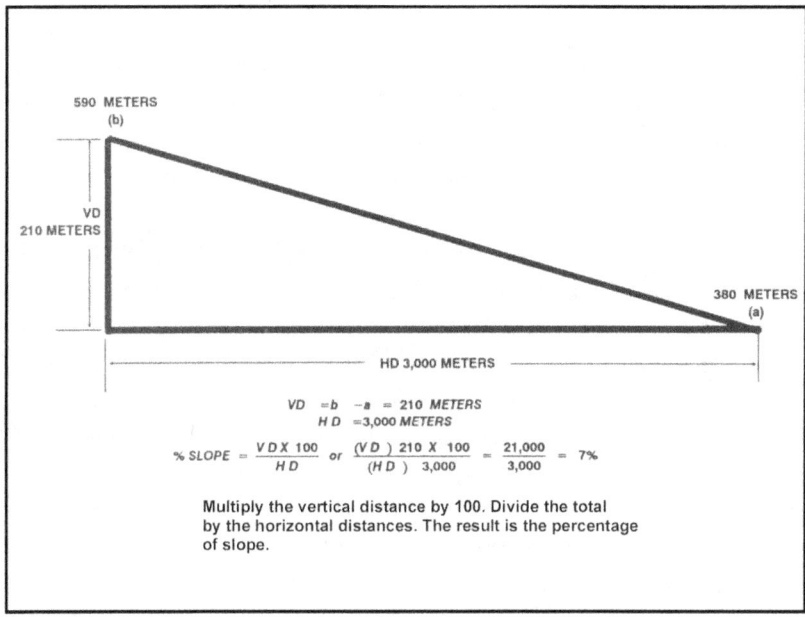

Figure 10-13. Percentage of slope in meters.

c. The slope angle can also be expressed in degrees. To do this, determine the VD and HD of the slope. Multiply the VD by 57.3 and then divide the total by the HD (Figure 10-14, page 10-10). This method determines the approximate degree of slope and is reasonably accurate for slope angles less than 20 degrees.

FM 3-25.26

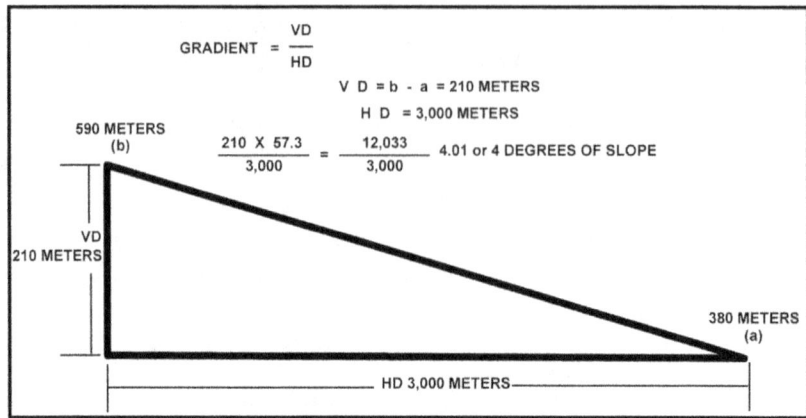

Figure 10-14. Degree of slope.

d. The slope angle can also be expressed as a gradient. The relationship of horizontal and vertical distance is expressed as a fraction with a numerator of one (Figure 10-15).

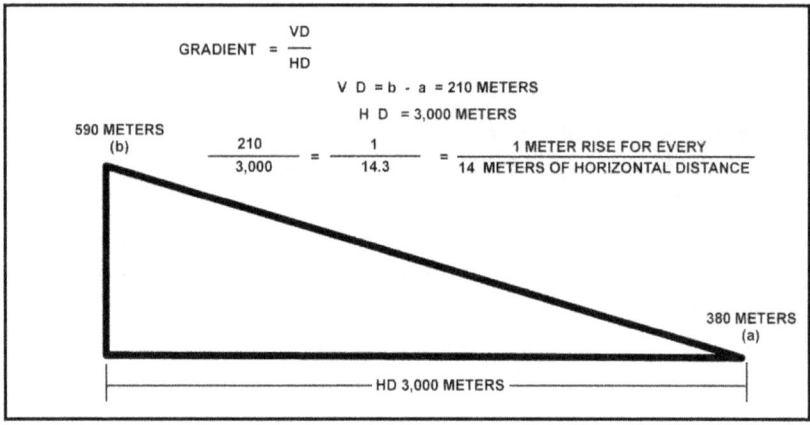

Figure 10-15. Gradient.

10-6. TERRAIN FEATURES

All terrain features are derived from a complex landmass known as a mountain or ridgeline (Figure 10-16). The term ridgeline is not interchangeable with the term ridge. A ridgeline is a line of high ground, usually with changes in elevation along its top and low ground on all sides from which a total of 10 natural or man-made terrain features are classified.

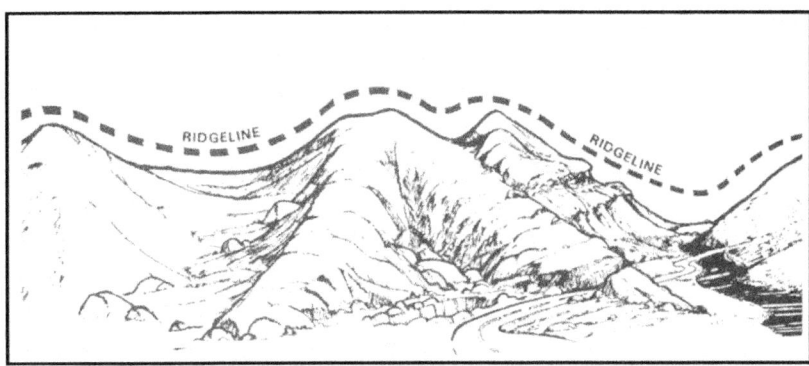

Figure 10-16. Ridgeline.

a. **Major Terrain Features.** Major terrain features are hills, saddles, valleys, ridges, and depressions.

(1) *Hill.* A hill is an area of high ground. From a hilltop, the ground slopes down in all directions. A hill is shown on a map by contour lines forming concentric circles. The inside of the smallest closed circle is the hilltop (Figure 10-17).

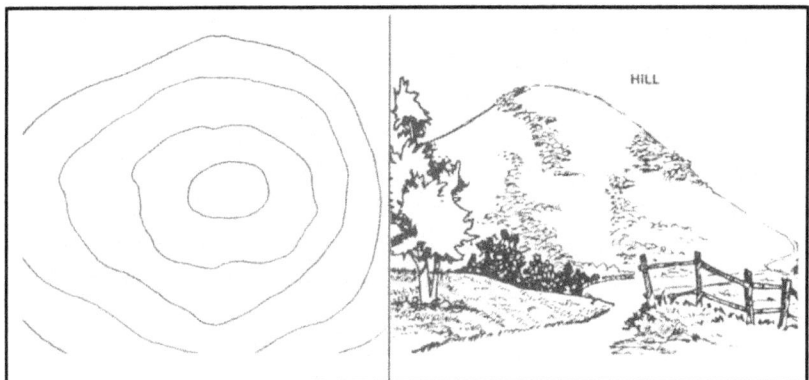

Figure 10-17. Hill.

(2) *Saddle.* A saddle is a dip or low point between two areas of higher ground. A saddle is not necessarily the lower ground between two hilltops; it may be simply a dip or break along a level ridge crest. If you are in a saddle, there is high ground in two opposite directions and lower ground in the other two directions. A saddle is normally represented as an hourglass (Figure 10-18, page 10-12).

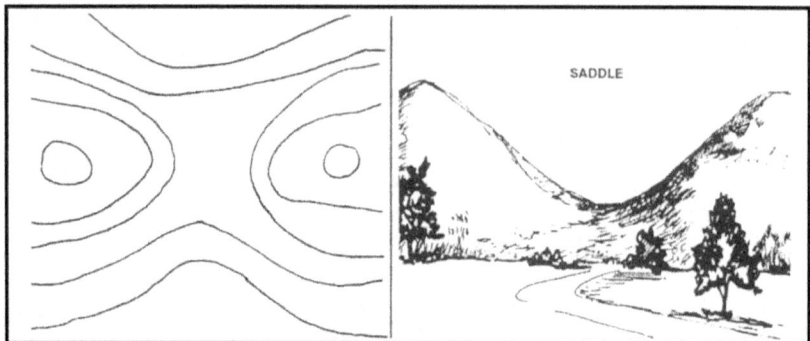

Figure 10-18. Saddle.

(3) ***Valley***. A valley is a stretched-out groove in the land, usually formed by streams or rivers. A valley begins with high ground on three sides and usually has a course of running water through it. If standing in a valley, three directions offer high ground, while the fourth direction offers low ground. Depending on its size and where a person is standing, it may not be obvious that there is high ground in the third direction, but water flows from higher to lower ground. Contour lines forming a valley are either U-shaped or V-shaped. To determine the direction water is flowing, look at the contour lines. The closed end of the contour line (U or V) always points upstream or toward high ground (Figure 10-19).

Figure 10-19. Valley.

(4) ***Ridge***. A ridge is a sloping line of high ground. If you are standing on the centerline of a ridge, you will normally have low ground in three directions and high ground in one direction with varying degrees of slope. If you cross a ridge at right angles, you will climb steeply to the crest and then descend steeply to the base. When you move along the path of the ridge, depending on the geographic location, there may be either an almost unnoticeable slope or a very obvious incline. Contour lines forming a ridge tend to be U-shaped or V-shaped. The closed end of the contour line points away from high ground (Figure 10-20).

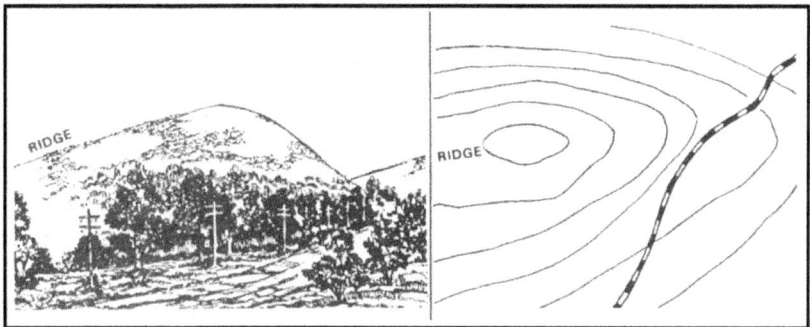

Figure 10-20. Ridge.

(5) *Depression.* A depression is a low point in the ground or a sinkhole. It could be described as an area of low ground surrounded by higher ground in all directions, or simply a hole in the ground. Usually only depressions that are equal to or greater than the contour interval will be shown. On maps, depressions are represented by closed contour lines that have tick marks pointing toward low ground (Figure 10-21).

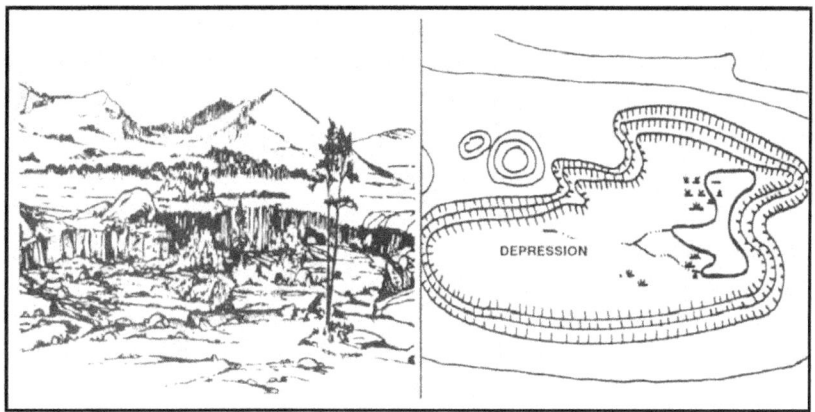

Figure 10-21. Depression.

b. **Minor Terrain Features.** Minor terrain features include draws, spurs, and cliffs.

(1) *Draw.* A draw is a stream course that is less developed than a valley. In a draw, there is essentially no level ground and, therefore, little or no maneuver room within its confines. If you are standing in a draw, the ground slopes upward in three directions and downward in the other direction. A draw could be considered as the initial formation of a valley. The contour lines depicting a draw are U-shaped or V-shaped, pointing toward high ground (Figure 10-22).

FM 3-25.26

Figure 10-22. Draw.

(2) **Spur**. A spur is a short, continuous sloping line of higher ground, normally jutting out from the side of a ridge. A spur is often formed by two roughly parallel streams cutting draws down the side of a ridge. The ground will slope down in three directions and up in one. Contour lines on a map depict a spur with the U or V pointing away from high ground (Figure 10-23).

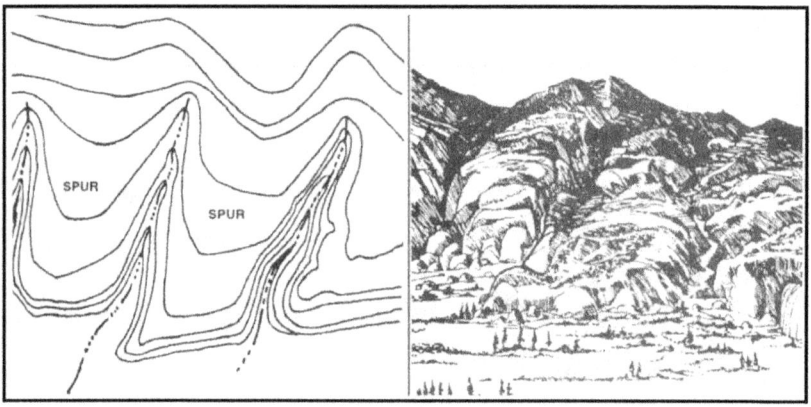

Figure 10-23. Spur.

(3) **Cliff**. A cliff is a vertical or near vertical feature; it is an abrupt change of the land. When a slope is so steep that the contour lines converge into one "carrying" contour of contours, this last contour line has tick marks pointing toward low ground (A, Figure 10-24). Cliffs are also shown by contour lines very close together and, in some instances, touching each other (B, Figure 10-24).

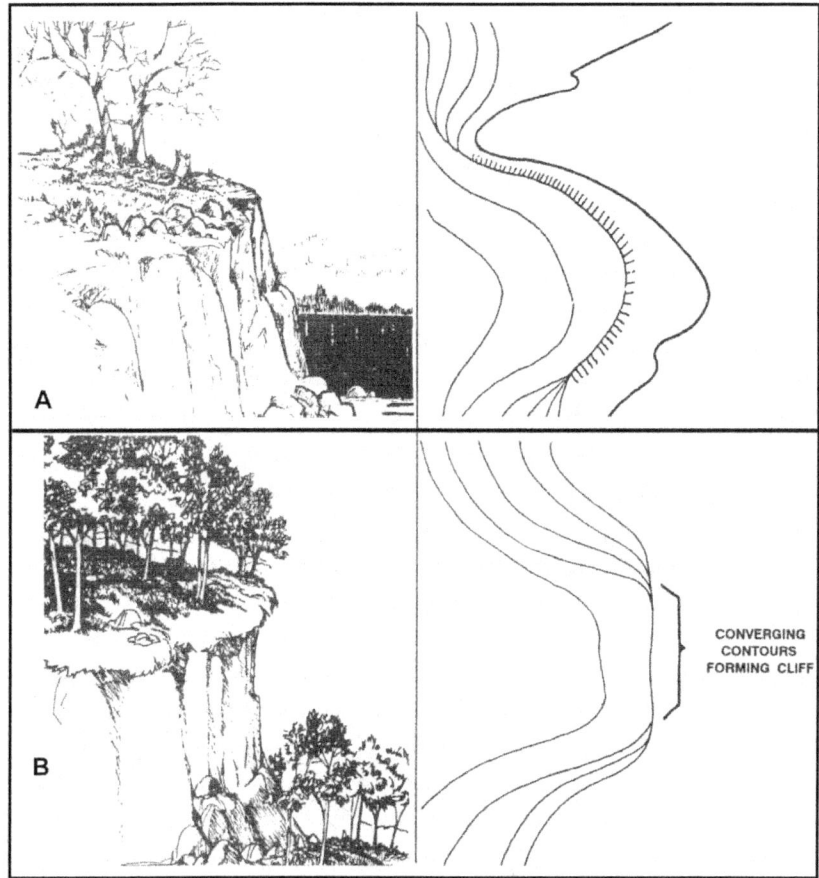

Figure 10-24. Cliff.

c. **Supplementary Terrain Features.** Supplementary terrain features include cuts and fills.

(1) *Cut.* A cut is a man-made feature resulting from cutting through raised ground, usually to form a level bed for a road or railroad track. Cuts are shown on a map when they are at least 10 feet high, and they are drawn with a contour line along the cut line. This contour line extends the length of the cut and has tick marks that extend from the cut line to the roadbed, if the map scale permits this level of detail (Figure 10-25, page 10-16).

(2) *Fill.* A fill is a man-made feature resulting from filling a low area, usually to form a level bed for a road or railroad track. Fills are shown on a map when they are at least 10 feet high, and they are drawn with a contour line along the fill line. This contour line extends the length of the filled area and has tick marks that point toward lower ground. If the map scale permits, the length of the fill tick marks are drawn to scale and extend from the base line of the fill symbol (Figure 10-25).

FM 3-25.26

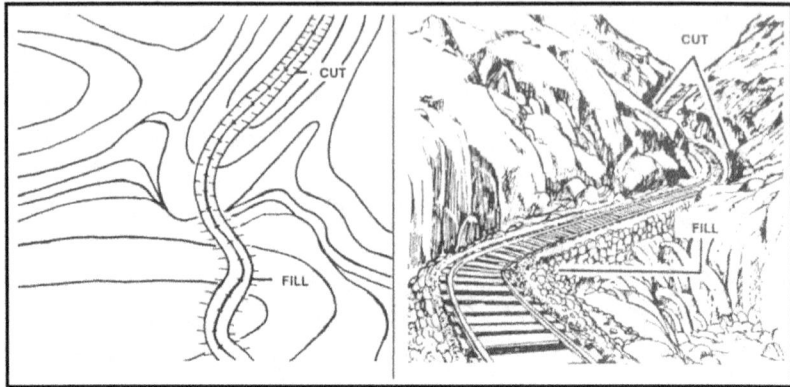

Figure 10-25. Cut and fill.

10-7. INTERPRETATION OF TERRAIN FEATURES

Terrain features do not normally stand alone. To better understand these when they are depicted on a map, you need to interpret them. You can interpret terrain features (Figure 10-26) by using contour lines; the shape, orientation, size, elevation, and slope (SOSES) approach; ridgelining; or streamlining.

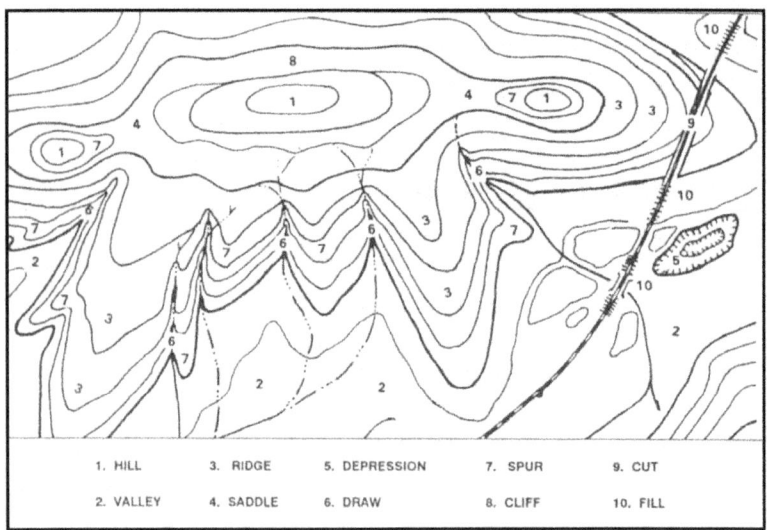

Figure 10-26. Terrain features.

a. **Contour Lines**. Emphasizing the main contour lines is a technique used to interpret the terrain of an area. By studying these contour lines, you can better understand the layout of the terrain and decide on the best route.

10-16 FOUO

18 January 2005

(1) The following description pertains to Figure 10-27 (page 10-19). Running east to west across the complex landmass is a ridgeline. A ridgeline is a line of high ground, usually with changes in elevation along its top and low ground on all sides. The changes in elevation are the three hilltops and two saddles along the ridgeline. From the top of each hill, there is lower ground in all directions. The saddles have lower ground in two directions and high ground in the opposite two directions. The contour lines of each saddle form half an hourglass shape. Because of the difference in size of the higher ground on the two opposite sides of a saddle, a full hourglass shape of a saddle may not be apparent.

(2) There are four prominent ridges. A ridge is on each end of the ridgeline and two ridges extend south from the ridgeline. All of the ridges have lower ground in three directions and higher ground in one direction. The closed ends of the U's formed by the contour lines point away from higher ground.

(3) To the south lies a valley; the valley slopes downward from east to west. Note that the U of the contour line points to the east, indicating higher ground in that direction and lower ground to the west. Another look at the valley shows high ground to the north and south of the valley.

(4) Just east of the valley is a depression. Looking from the bottom of the depression, there is higher ground in all directions.

(5) There are several spurs extending generally south from the ridgeline. They, like ridges, have lower ground in three directions and higher ground in one direction. Their contour line U's point away from higher ground.

(6) Between the ridges and spurs are draws. They, like valleys, have higher ground in three directions and lower ground in one direction. Their contour line U's and V's point toward higher ground.

(7) Two contour lines on the north side of the center hill are touching or almost touching. They have ticks indicating a vertical or nearly vertical slope or a cliff.

(8) The road cutting through the eastern ridge depicts cuts and fills. The breaks in the contour lines indicate cuts, and the ticks pointing away from the road bed on each side of the road indicate fills.

b. **SOSES.** A recommended technique for identifying specific terrain features and then locating them on the map is to use five characteristics known by the mnemonic SOSES. Terrain features can be examined, described, and compared with each other and with corresponding map contour patterns in terms of their shapes, orientations, sizes, elevations, and slopes. Through practice, you can learn to identify several individual terrain features in the field and see how they vary in appearance.

(1) *Shape.* Shape is the general form or outline of the feature at its base.

(2) *Orientation.* Orientation is the general trend or direction of a feature from your viewpoint. A feature can be in line, across, or at an angle to your viewpoint.

(3) *Size.* Size is the length or width of a feature horizontally across its base. For example, one terrain feature might be larger or smaller than another.

(4) *Elevation.* Elevation is the height of a terrain feature. This can be described either in absolute or relative terms as compared to the other features in the area. One landform may be higher, lower, deeper, or shallower than another.

(5) *Slope.* Slope is the type (uniform, convex, or concave) and steepness or angle (steep or gentle) of the sides of a terrain feature.

NOTE: Further terrain analysis using SOSES can be learned by taking the Map Interpretation and Terrain Association Course. It consists of three separate courses of instruction: basic, intermediate, and advanced. Using photographic slides of terrain and other features, basic instruction teaches how to identify basic terrain feature types on the ground and on the map. Intermediate instruction teaches elementary map interpretation and terrain association using real world scenes and map sections of the same terrain. Advanced instruction teaches advanced techniques for map interpretation and terrain association. The primary emphasis is on the concepts of map design guidelines and terrain association skills. Map design guidelines refer to the rules and practices used by cartographers in the compilation and symbolization of military topographic maps. Knowledge of the selection, classification, and symbolization of mapped features greatly enhances the user's ability to interpret map information.

c. **Ridgelining.** This technique helps you to visualize the overall lay of the ground within the area of interest on the map. Use the following steps to implement this technique.

(1) Identify on the map the crests of the ridgelines in your area of operation by identifying the close-out contours that lie along the hilltop.

(2) Trace over the crests so each ridgeline stands out clearly as one identifiable line. The usual colors used for this tracing are red or brown; however, you may use any color at hand.

(3) Go back over each of the major ridgelines and trace over the prominent ridges and spurs that come out of the ridgelines.

(4) When you have completed the ridgelining process, you will find that the high ground on the map will stand out and that you will be able to see the relationship between the various ridgelines (Figure 10-27).

d. **Streamlining.** This procedure (Figure 10-27) is similar to that of ridgelining.

(1) Identify all the mapped streams in the area of operations.

(2) Trace over them to make them stand out more prominently. The color used for this is usually blue; but again, if blue is not available, use any color at hand so long as the distinction between the ridgelines and the streamlines is clear.

(3) Identify other low ground, such as smaller valleys or draws that feed into the major streams, and trace over them. This brings out the drainage pattern and low ground in the area of operation on the map.

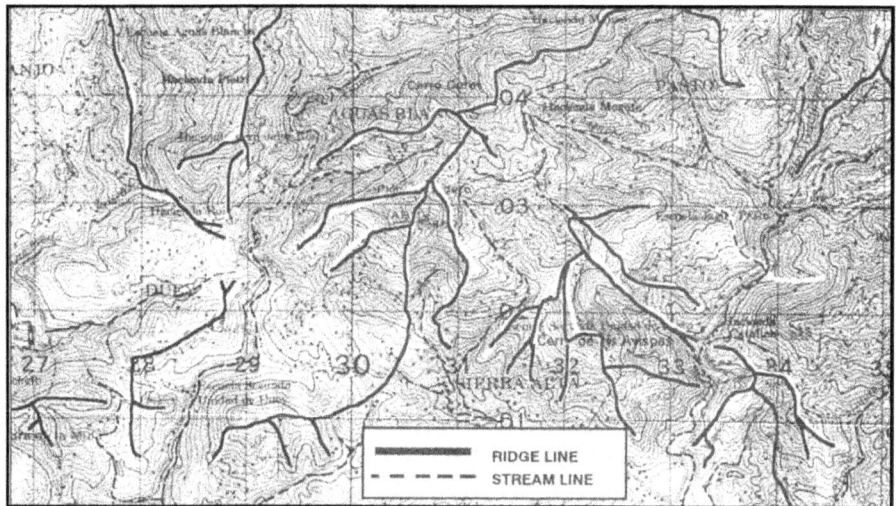

Figure 10-27. Ridgelining and streamlining.

10-8. PROFILES

The study of contour lines to determine high and low points of elevation is usually adequate for military operations. However, there may be times when a quick and precise reference to determine exact elevations of specific points is needed. When exactness is demanded, a profile is required. A profile, within the scope and purpose of this manual, is an exaggerated side view of a portion of the earth's surface along a line between two or more points.

a. A profile can be used for many purposes. The primary purpose is to determine if line of sight is available. Line of sight is used—

- To determine defilade positions.
- To plot hidden areas or dead space.
- To determine potential direct fire weapon positions.
- To determine potential locations for defensive positions.
- To conduct preliminary planning in locating roads, pipelines, railroads, or other construction projects.

b. A profile can be constructed from any contoured map using the following steps:

(1) Draw a line on the map from where the profile is to begin to where it is to end (Figure 10-28, page 10-20).

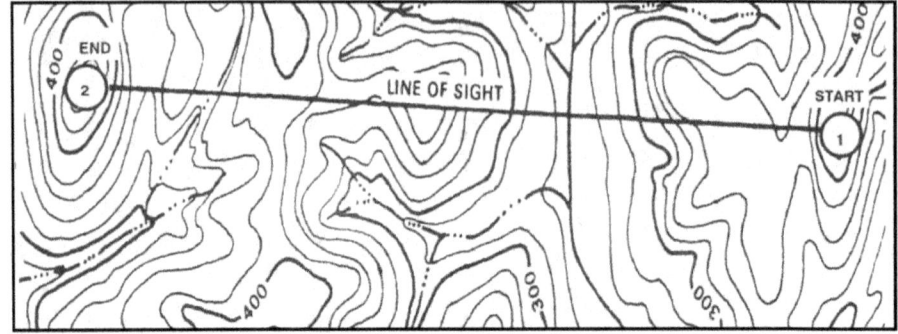

Figure 10-28. Connecting points.

(2) Find the value of the highest and lowest contour lines that cross or touch the profile line. Add one contour value above the highest and one below the lowest to take care of hills and valleys.

(3) Select a piece of lined notebook paper with as many lines as was determined in (2) above. The standard Army green pocket notebook or any other paper with 1/4-inch lines is ideal. Wider lines, up to 5/8-inch, may be used. If lined paper is not available, draw equally spaced horizontal lines on a blank sheet of paper.

(4) Number the top line with the highest value and the bottom line with the lowest value as determined in (2) above.

(5) Number the rest of the lines in sequence, starting with the second line from the top. The lines will be numbered in accordance with the contour interval (Figure 10-29).

(6) Place the paper on the map with the lines next to and parallel to the profile line (Figure 10-29).

(7) From every point on the profile line where a contour line, stream, intermittent stream, or other body of water crosses or touches, drop a perpendicular line to the line having the same value. Place a tick mark where the perpendicular line crosses the number line (Figure 10-29). Where trees are present, add the height of the trees to the contour line and place a tick mark there. Assume the height of the trees to be 50 feet or 15 meters where dark green tint is shown on the map. Vegetation height may be adjusted up or down when operations in the area have provided known tree heights.

(8) After all perpendicular lines have been drawn and tick marks placed where the lines cross, connect all tick marks with a smooth, natural curve to form a horizontal view or profile of the terrain along the profile line (Figure 10-29).

(9) The profile drawn may be exaggerated. The spacing between the lines drawn on the sheet of paper will determine the amount of exaggeration and may be varied to suit any purpose.

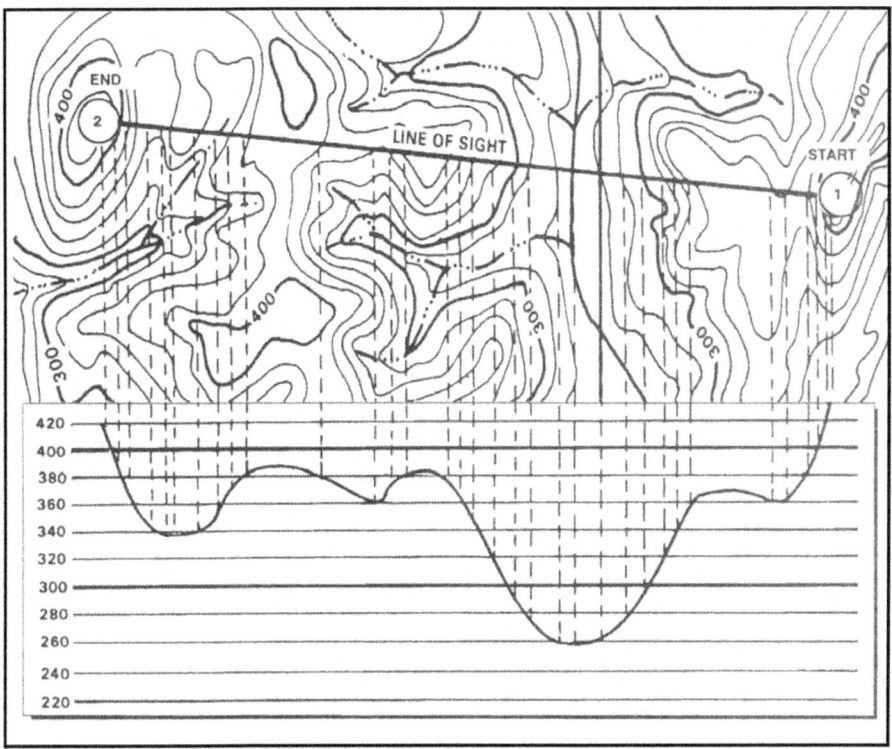

Figure 10-29. Dropping perpendiculars.

(10) Draw a straight line from the start point to the end point on the profile. If the straight line intersects the curved profile, line of sight to the end point is not available (Figure 10-30).

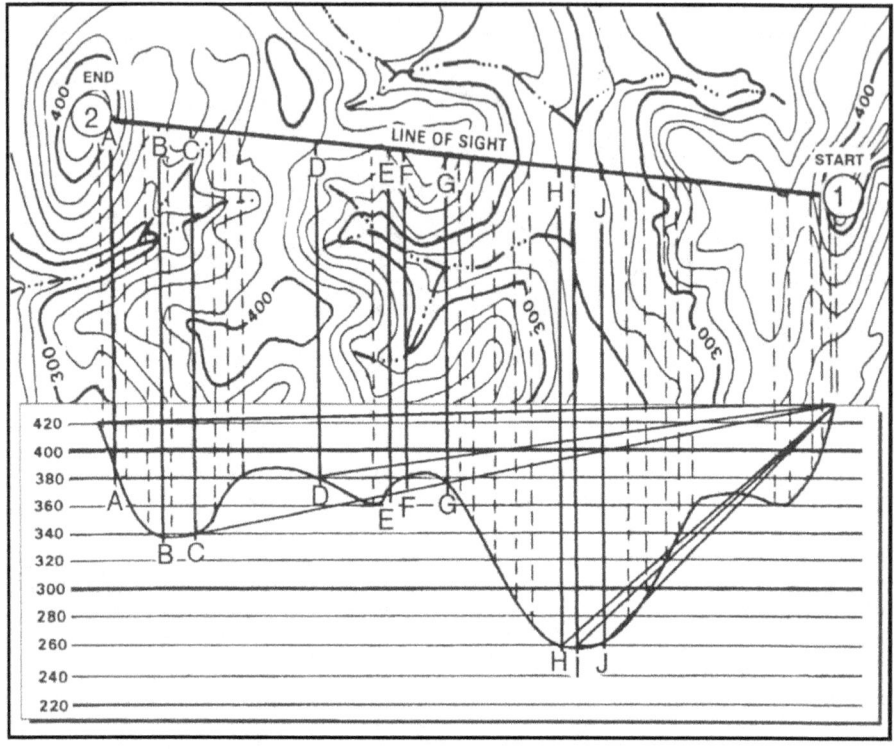

Figure 10-30. Drawing lines to additional points.

(11) Line of sight to other points along the profile line can be determined by drawing a line from the start point to additional points. In Figure 10-31 line of sight is available to—

A—Yes	D—Yes	G—Yes	J—No
B—No	E—No	H—No	
C—No	F—No	I—No	

(12) The vertical distance between navigable ground up to the line of sight line is the depth of defilade.

c. When time is short, or when a complete profile is not needed, one may be constructed showing only the hilltops, ridges, and if desired, the valleys. This is called a hasty profile. It is constructed in the same manner as a full profile (Figure 10-31).

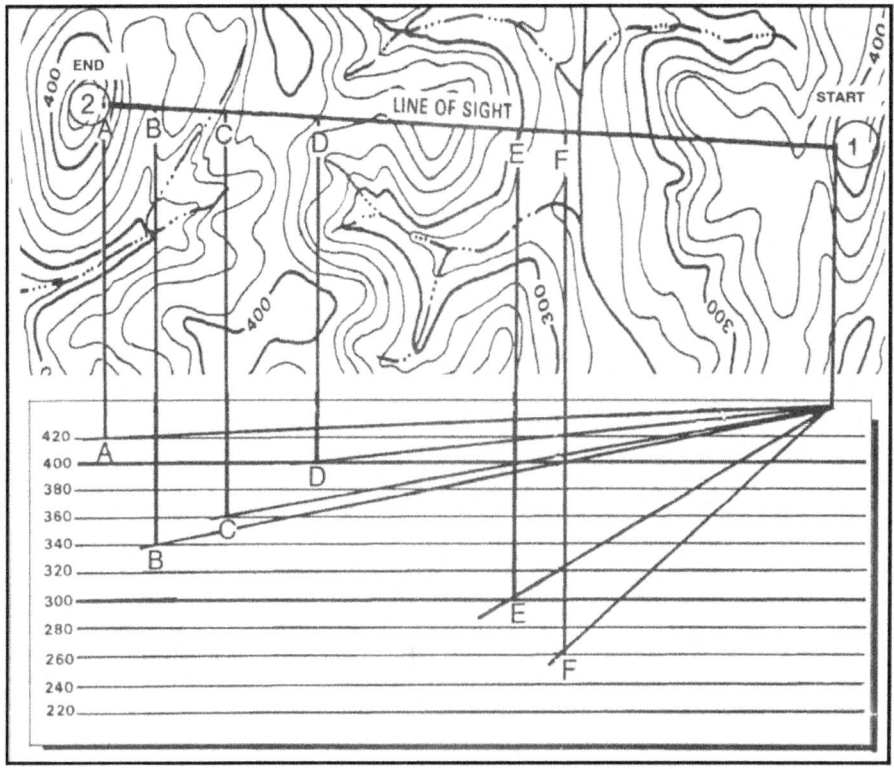

Figure 10-31. Drawing a hasty profile.

This Page intentionally left blank.

CHAPTER 11
TERRAIN ASSOCIATION

Failing to use the vast amounts of information presented by the map and available to the eye on the ground reduces the chances for success in land navigation. The Soldier who has repeatedly practiced the skills of identifying and discriminating among the many types of terrain and other features knows how these features are mapped. By studying the map, he can begin to visualize the shape of the land, estimate distances, and perform quick resection from the many landmarks he sees. This Soldier is the one who will be at the right place to help defeat the enemy on the battlefield.

This chapter tells how to orient a map with and without a compass, how to find locations on a map as well as on the ground, how to study the terrain, and how to move on the ground using terrain association and dead reckoning.

11-1. ORIENTATION OF THE MAP

The first step for a navigator in the field is orienting the map. A map is oriented when it is in a horizontal position with its north and south corresponding to the north and south on the ground. Some orienting techniques are described herein.

   a. **Using a Compass.** When orienting a map with a compass, remember that the compass measures magnetic azimuths. Since the magnetic arrow points to magnetic north, pay special attention to the declination diagram. Two techniques are used.

   (1) *First Technique.* Determine the direction of the declination and its value from the declination diagram.

   (a) With the map in a horizontal position, take the straightedge on the left side of the compass and place it alongside the north-south grid line with the cover of the compass pointing toward the top of the map. This procedure places the fixed black index line of the compass parallel to north-south grid lines of the map.

   (b) Keeping the compass aligned as directed above, rotate the map and compass together until the magnetic arrow is below the fixed black index line on the compass. At this time, the map is close to being oriented.

   (c) Rotate the map and compass in the direction of the declination diagram.

   (d) If the magnetic north arrow on the map is to the left of the grid north, check the compass reading to see if it equals the G-M angle given in the declination diagram. The map is then oriented (Figure 11-1, page 11-2).

FM 3-25.26

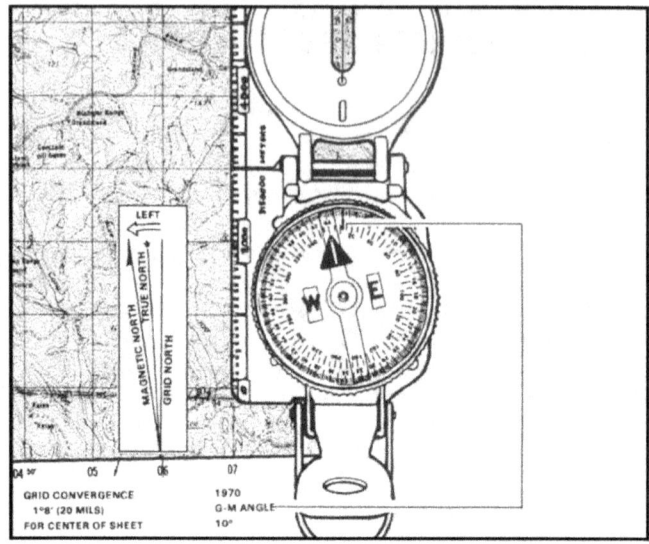

Figure 11-1. Map oriented with 10 degrees west declination.

(e) If the magnetic north is to the right of grid north, check the compass reading to see if it equals 360 degrees minus the G-M angle (Figure 11-2).

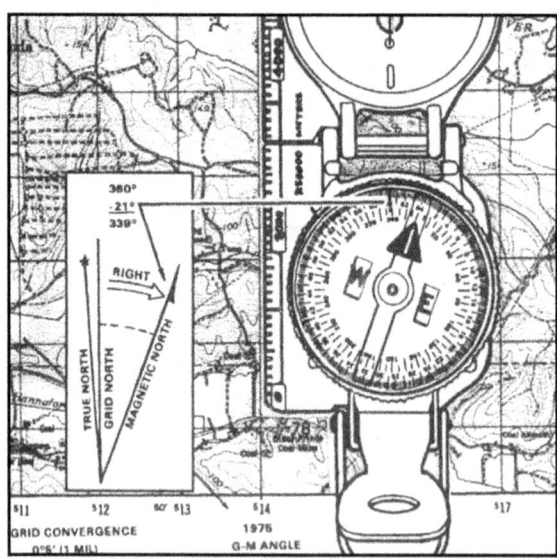

Figure 11-2. Map oriented with 21 degrees east declination.

(2) *Second Technique.* Determine the direction of the declination and its value from the declination diagram.

(a) Using any north-south grid line on the map as a base, draw a magnetic azimuth equal to the G-M angle given in the declination diagram with the protractor.

(b) If the declination is easterly (right), the drawn line is equal to the value of the G-M angle. Then align the straightedge on the left side of the compass alongside the drawn line on the map. Rotate the map and compass until the magnetic arrow of the compass is below the fixed black index line. The map is now oriented (Figure 11-3).

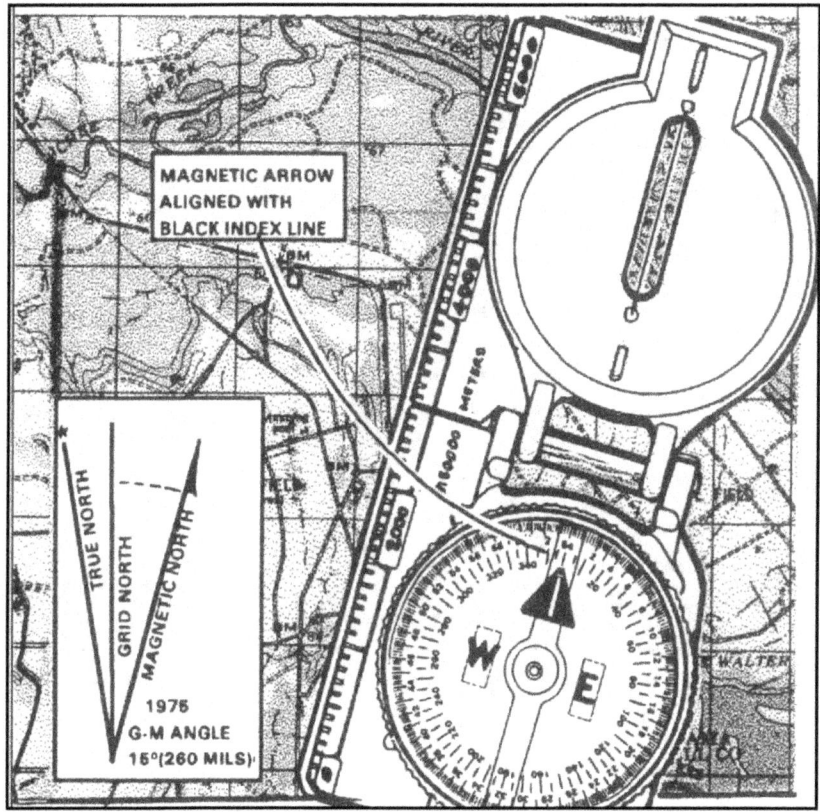

Figure 11-3. Map oriented with 15 degrees east declination.

(c) If the declination is westerly (left), the drawn line will equal 360 degrees minus the value of the G-M angle. Then align the straightedge on the left side of the compass alongside the drawn line on the map. Rotate the map and compass until the magnetic arrow of the compass is below the fixed black index line. The map is now oriented (Figure 11-4).

NOTES: 1. Once the map is oriented, magnetic azimuths are determined using the compass. Do not move the map from its oriented position since any change in

FM 3-25.26

its position moves it out of line with the magnetic north. [See paragraph 11-6b(1).]
2. Special care should be taken when orienting your map with a compass. A small mistake can cause you to navigate in the wrong direction.

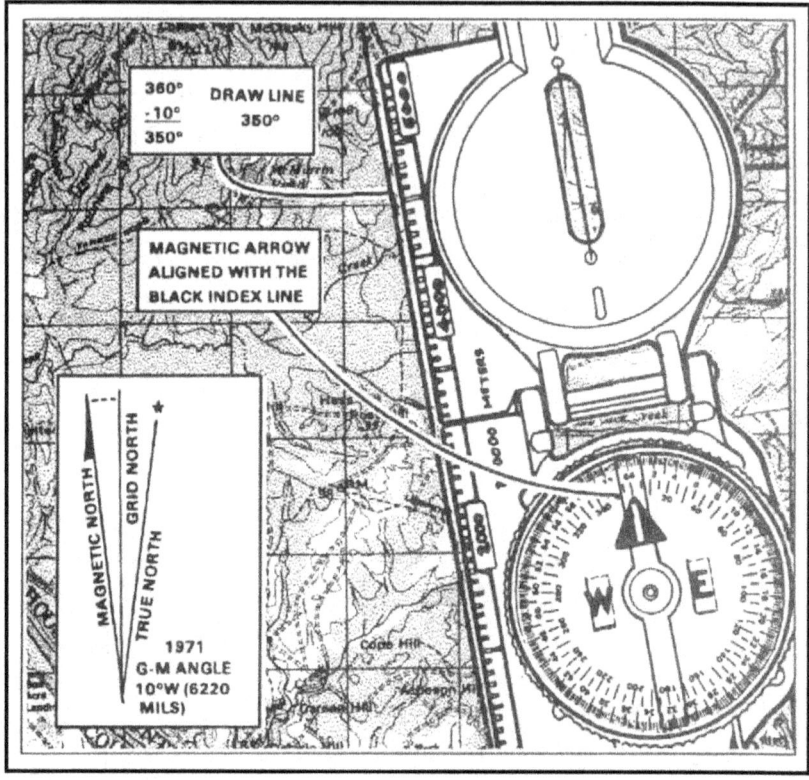

Figure 11-4. Map oriented with 10 degrees west declination.

b. **Using Terrain Association.** A map can be oriented by terrain association when a compass is not available or when the user has to make many quick references as he moves across country. Using this method requires careful examination of the map and the ground, and the user must know his approximate location (Figure 11-5). (Orienting by this method is discussed in detail in paragraph 11-3.)

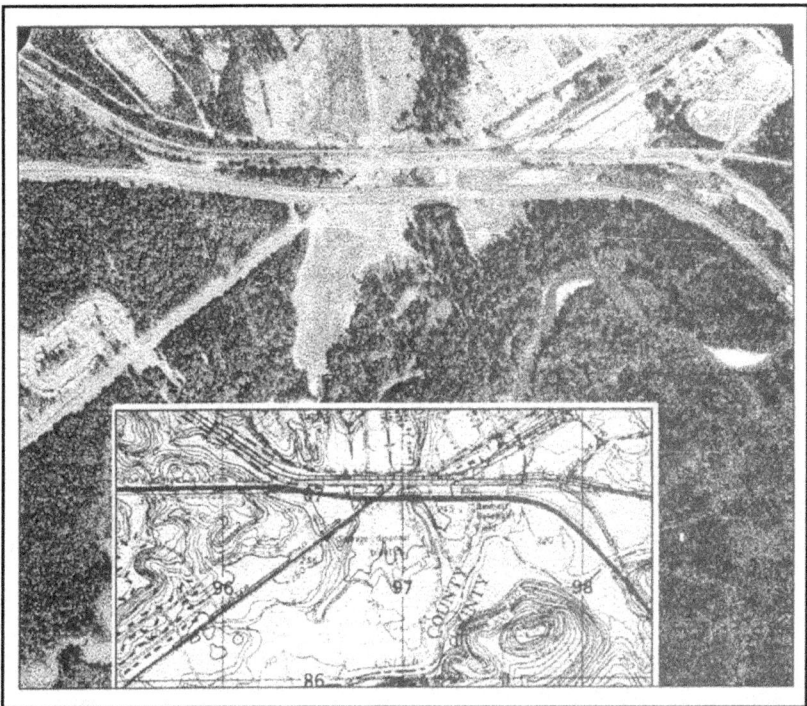

Figure 11-5. Terrain association.

FM 3-25.26

c. **Using Field-Expedient Methods.** When a compass is not available and there are no recognizable terrain features, a map may be oriented by any of the field-expedient methods described in Chapter 9, paragraph 9-5. (Also, see Figure 11-6.)

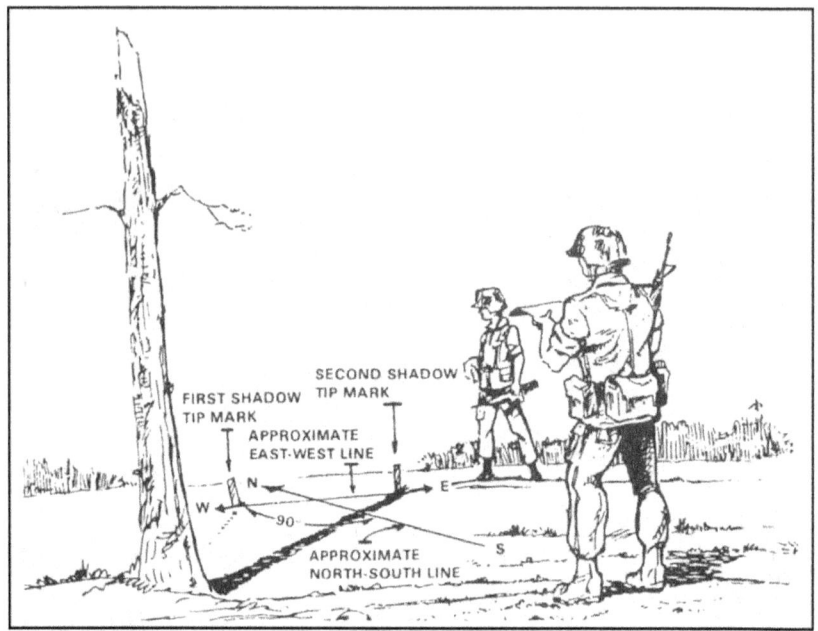

Figure 11-6. Field-expedient method.

11-2. LOCATIONS

The key to success in land navigation is to know your location at all times. With this basic knowledge, you can decide which direction and how far to travel.

   a. **Known Position.** Most important of all is the initial location of the user before starting any movement in the field. If movement takes place without establishing the initial location, everything that is done in the field from there on is a gamble. Determine the initial location by referring to the last known position, by grid coordinates and terrain association, or by locating and orienting your position on the map and ground.

   b. **Known Point/Known Distance (Polar Plot).** This location can be determined by knowing the starting point, the azimuth to the desired objective, and the distance to it.

   c. **Resection, Modified Resection, and Intersection.** See Chapter 6.

   d. **Indirect Fire.** Finding a location by indirect fire is done with smoke. Use the point of impact of the round as a reference point from which distances and azimuth can be obtained.

11-3. TERRAIN ASSOCIATION USAGE

The technique of moving by terrain association is more forgiving of mistakes and far less time-consuming than dead reckoning. It best suits those situations that call for movement from one area to another. Errors made using terrain association are easily corrected because you are comparing what you expected to see from the map to what you do see on the ground. Errors are anticipated and will not go unchecked. You can easily make adjustments based upon what you encounter. Periodic position-fixing through either plotted or estimated resection will also make it possible to correct your movements, call for fire, or call in the locations of enemy targets or any other information of tactical or logistical importance.

   a. **Matching the Terrain to the Map by Examining Terrain Features.** By observing the contour lines in detail, the five major terrain features (hilltop, valley, ridge, depression, and saddle) should be determined. This is a simple task in an area where the observer has ample view of the terrain in all directions. One-by-one, match the terrain features depicted on the map with the same features on the ground. In restricted terrain, this procedure becomes harder; however, constantly check the map as you move since it is the determining factor (Figure 11-5).

   b. **Comparing the Vegetation Depicted on the Map.** When comparing the vegetation, a topographic map should be used to make a comparison of the clearings that appear on the map with the ones on the ground. The user must be familiar with the different symbols, such as vineyards, plantations, and orchards, that appear on the legend. The age of the map is an important factor when comparing vegetation. Some important vegetation features were likely to be different when the map was made. Another important factor about vegetation is that it can change overnight by natural accidents or by man (forest fires, clearing of land for new developments, farming, and so forth).

   c. **Masking by the Vegetation.** Camouflage the important landforms using vegetation. Use of camouflage makes it harder for the navigator to use terrain association.

   d. **Using the Hydrography.** Inland bodies of water can help during terrain association. The shape and size of lakes in conjunction with the size and direction of flow of the rivers and streams are valuable help.

   e. **Using Man-made Features.** Man-made features are an important factor during terrain association. The user must be familiar with the symbols shown in the legend representing those features. The direction of buildings, roads, bridges, high-tension lines, and so forth make the terrain inspection a lot easier; however, the age of the map must be considered because man-made features appear and disappear constantly.

   f. **Examining the Same Piece of Terrain During the Different Seasons of the Year.** In those areas of the world where the seasons are distinctive, a detailed examination of the terrain should be made during each season. The same piece of land does not present the same characteristics during both spring and winter.

   (1) During winter, the snow packs the vegetation, delineating the land, making the terrain features appear as clear as they are shown by the contour lines on the map. Ridges, valleys, and saddles are very distinctive.

   (2) During spring, the vegetation begins to reappear and grow. New vegetation causes a gradual change of the land to the point that the foliage conceals the terrain features and makes the terrain hard to recognize.

   (3) During summer months, the effects are similar to those in the spring.

   (4) Fall makes the land appear different with its change of color and gradual loss of vegetation.

   (5) During the rainy season, the vegetation is green and thick, and the streams and ponds look like small rivers and lakes. In sparsely vegetated areas, the erosion changes the shape of the land.

(6) During a period of drought, the vegetation dries out and becomes vulnerable to forest fires that change the terrain whenever they occur. Also during this season, the water levels of streams and lakes drop, adding new dimensions and shape to the existing mapped areas.

g. **Following an Example of Terrain Association.** Your location is hilltop 514 in the left center of the map in Figure 11-7.

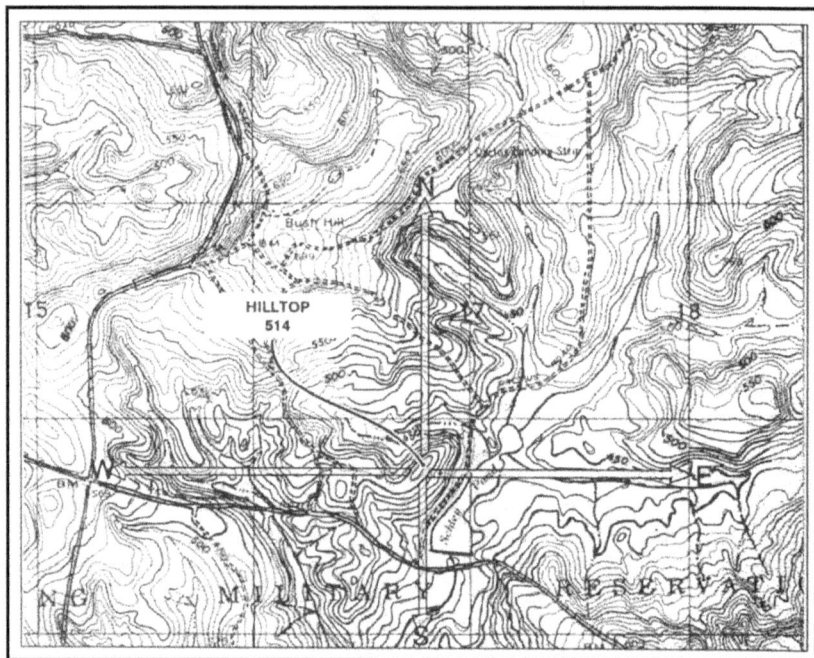

Figure 11-7. Example of terrain association.

(1) *To The North*. The contour lines indicate that the hill slopes down for about 190 meters, and that it leads into a small valley containing an intermittent stream. On the other side of the stream as you continue with your northerly inspection, the terrain starts a gradual ascent, indicating a hilltop partially covered with vegetation, until an unimproved road is reached. This road runs along a gradual ridgeline with northwest direction. Then the contour line spacings become narrow, indicating a steeper grade that leads to a narrow valley containing a small intermittent stream. As you continue up, you find a small but prominent ridge with a clearing. The contour lines once again show a steeper grade leading to a moderate valley containing an intermittent stream running in a southeast direction.

(2) *To The East*. There is a clearing of the terrain as it slopes down to Schley Pond. An ample valley is clearly seen on the right side of the pond, as indicated by the "U" and "V" shape of the contour lines. This valley contains some swamp areas and there is a long ridgeline on the north portion of the valley.

(3) *To The South*. The terrain gently slopes downward until a clear area is reached. It continues in a downward direction to an intermittent stream running southeast in a small

valley. There is also an improved road running in the same direction as the valley. At the intersection of the roads as you face south, there is a clearing of about 120 meters on the ridge. At the bottom of it, a stream runs from Schley Pond in a southwest direction through an ample valley fed by two intermittent streams. As you continue, a steep, vegetated hill is found with a clearing on its top, followed by a small saddle and another hilltop.

(4) *To The West.* First, you see a small, clear valley. It is followed by a general ridgeline running northwest in which an unimproved road is located just before a hilltop. Continuing on a westerly direction, you will find a series of alternate valleys and ridges.

11-4. TACTICAL CONSIDERATIONS

Military cross-country navigation is intellectually demanding because it is imperative that the unit, crew, or vehicle survive and successfully complete the move in order to accomplish its mission. However, the unnecessary use of a difficult route makes navigation too complicated, creates more noise when proceeding over it, causes wear and tear on equipment and personnel, increases the need for and needlessly complicates recovery operations, and wastes scarce time. On receipt of a tactical mission, the leader begins his troop-leading procedures and makes a tentative plan. He bases the tentative plan on a good terrain analysis. He analyzes the considerations covered in the following mnemonics—OCOKA and METT-TC.

a. **OCOKA.** The terrain should be analyzed for observation and fields of fire, cover and concealment, obstacles, key terrain, and avenues of approach.

(1) *Observation and Fields of Fire.* The purpose of observation is to see the enemy (or various landmarks) but not be seen by him. Anything that can be seen can be hit. Therefore, a field of fire is an area that a weapon or a group of weapons can cover effectively with fire from a given position.

(2) *Cover and Concealment.* Cover is shelter or protection (from enemy fire) either natural or artificial. Always try to use covered routes and seek cover for each halt, no matter how brief it is planned to be. Unfortunately, two factors interfere with obtaining constant cover. One is time and the other is terrain. Concealment is protection from observation or surveillance, including concealment from enemy air observation. Before, trees provided good concealment, but with modern thermal and infrared imaging equipment, trees are not always effective. When you are moving, concealment is generally secondary; therefore, select routes and positions that do not allow covered or concealed enemy near you.

(3) *Obstacles.* Obstacles are any obstructions that stop, delay, or divert movement. Obstacles can be natural (rivers, swamps, cliffs, or mountains) or they may be artificial (barbed wire entanglements, pits, concrete or metal antimechanized traps). They can be ready-made or constructed in the field. Always consider any possible obstacles along your movement route and, if possible, try to keep obstacles between the enemy and yourself.

(4) *Key Terrain.* Key terrain is any locality or area that the seizure or retention of affords a marked advantage to either combatant. Urban areas are often seen by higher headquarters as being key terrain because they are used to control routes. On the other hand, an urban area that is destroyed may be an obstacle instead. High ground can be key because it dominates an area with good observation and fields of fire. In an open area, a draw or wadi (dry streambed located in an arid area) may provide the only cover for many kilometers, thereby becoming key. You should always attempt to locate any area near you that could be even remotely considered as key terrain.

(5) ***Avenues of Approach.*** These are access routes. They may be the routes you can use to get to the enemy or the routes they can use to get to you. Basically, an identifiable route that approaches a position or location is an avenue of approach to that location. They are often terrain corridors such as valleys or wide, open areas.

b. **METT-TC.** Tactical factors other than the military aspects of terrain must also be considered in conjunction with terrain during movement planning and execution as well. These additional considerations are mission, enemy, terrain and weather, troops, and time available.

(1) ***Mission.*** This refers to the specific task assigned to a unit or individual. It is the duty or task together with the purpose that clearly indicates the action to be taken and the reason for it—but not how to do it. Training exercises should stress the importance of a thorough map reconnaissance to evaluate the terrain. This allows the leader to confirm his tentative plan, basing his decision on the terrain's effect on his mission.

(a) Marches by foot or vehicle are used to move troops from one location to another. Soldiers must get to the right place, at the right time, and in good fighting condition. The normal rate for an 8-hour foot march is 4 kilometers per hour. However, the rate of march may vary, depending on the following factors:

- Distance.
- Time allowed.
- Likelihood of enemy contact.
- Terrain.
- Weather.
- Physical condition of Soldiers.
- Equipment/weight to be carried.

A motor march requires little or no walking by the Soldiers, but the factors affecting the rate of march still apply.

(b) Patrol missions are used to conduct combat or reconnaissance operations. Without detailed planning and a thorough map reconnaissance, any patrol mission may not succeed. During the map reconnaissance, the mission leader determines a primary and alternate route to and from the objectives.

(c) Movement to contact is conducted whenever an element is moving toward the enemy but is not in contact with the enemy. The lead element must orient its movement on the objective by conducting a map reconnaissance, determining the location of the objective on both the map and the ground, and selecting the route to be taken.

(d) Delays and withdrawals are conducted to slow the enemy down without becoming decisively engaged, or to assume another mission. To be effective, the element leader must know where he is to move and the route to be taken.

(2) ***Enemy.*** This refers to the strength, status of training, disposition (locations), doctrine, capabilities, equipment (including night vision devices), and probable courses of action that impact upon both the planning and execution of the mission, including a movement.

(3) ***Terrain and Weather.*** Observation and fields of fire influence the placement of positions and crew-served weapons. The leader conducts a map reconnaissance to determine key terrain, obstacles, cover and concealment, and likely avenues of approach.

(a) Key terrain is any area whose control affords a marked advantage to the force holding it. Some types of key terrain are high ground, bridges, towns, and road junctions.

(b) Obstacles are natural or man-made terrain features that stop, slow down, or divert movement. Consideration of obstacles is influenced by the unit's mission. An obstacle may be an advantage or disadvantage, depending upon the direction of attack or defense. Obstacles can be found by conducting a thorough map reconnaissance and study of recent aerial photographs.

(c) Cover and concealment are determined for both friendly and enemy forces. Concealment is protection from observation; cover is protection from the effects of fire. Most terrain features that offer cover also provide concealment from ground observation. There are areas that provide no concealment from enemy observation. These danger areas may be large or small open fields, roads, or streams. During the leader's map reconnaissance, he determines any obvious danger areas and, if possible, adjusts his route.

(d) Avenues of approach (AAs) are routes by which a unit may reach an objective or key terrain. To be considered an AA, a route must provide enough width for the deployment of the size force for which it is being considered. The AAs are also considered for the subordinate enemy force. For example, a company determines likely AAs for an enemy platoon; a platoon determines likely AAs for an enemy squad. Likely AAs may be either ridges, valleys, or by air. By examining the terrain, the leader determines the likely enemy AAs based on the tactical situation.

(e) Weather has little effect on dismounted land navigation. Rain and snow could possibly slow down the rate of march, that is all. But during mounted land navigation, the navigator must know the effect of weather on his vehicle. (See Chapter 12 for mounted land navigation.)

(4) **Troops**. Consideration of your own troops is equally important. The size and type of the unit to be moved and its capabilities, physical condition, status of training, and types of equipment assigned all affect the selection of routes, positions, fire plans, and the various decisions to be made during movement. On ideal terrain such as relatively level ground with little or no woods, a platoon can defend a front of up to 400 meters. The leader must conduct a thorough map reconnaissance and terrain analysis of the area his unit is to defend. Heavily wooded areas or very hilly areas may reduce the front a platoon can defend. The size of the unit must also be taken into consideration when planning a movement to contact. During movement, the unit must retain its ability to maneuver. A small draw or stream may reduce the unit's maneuverability but provide excellent concealment. All of these factors must be considered.

(a) Types of equipment that may be needed by the unit can be determined by a map reconnaissance. For example, if the unit must cross a large stream during its movement to the objective, ropes may be needed for safety lines.

(b) Physical capabilities of the Soldiers must be considered when selecting a route. Crossing a large swampy area may present no problem to a physically fit unit, but to a unit that has not been physically conditioned, the swampy area may slow or completely stop its movement.

(5) **Time Available**. At times, the unit may have little time to reach an objective or to move from one point to another. The leader must conduct a map reconnaissance to determine the quickest route to the objective; this is not always a straight route. From point A to point B on the map may appear to be 1,000 meters, but if the route is across a large ridge, the distance will be greater. Another route from point A to B may be 1,500 meters—but on flat

terrain. In this case, the quickest route would be across the flat terrain; however, concealment and cover may be lost.

(6) **Civil Considerations.** Civil considerations are present throughout offensive operations. They may preclude the attack of some targets, such as infrastructure and historically significant areas, and may limit the use of land mines.

(a) Commanders focus their staffs on considerations that may affect mission accomplishment. These factors include care and support for civilians within the AO and the possible effect of refugees on operations and movements. Other considerations include enemy locations with respect to civil populations, political and cultural boundaries, and language requirements.

(b) Enemy propaganda may affect the attitude of civilians in the AO. It may also affect domestic and foreign support for the operation. Operations commanders pay particular attention to the effects of actions in the information environment. Tactical commanders may have limited awareness of media reporting and its effect on public opinion. Operational commanders gauge the effect of public opinion and keep their subordinates informed.

11-5. MOVEMENT AND ROUTE SELECTION

One key to success in tactical missions is the ability to move undetected to the objective. There are four steps to land navigation. Being given an objective and the requirement to move there, you must know where you are, plan the route, stay on the route, and recognize the objective.

    a. **Know Where You Are (Step 1).** You must know where you are on the map and on the ground at all times and in every possible way. This includes knowing where you are relative to—

- Your directional orientation.
- The direction and distances to your objective.
- Other landmarks and features.
- Any impassable terrain, the enemy, and danger areas.
- Both the advantages and disadvantages presented by the terrain between you and your objective.

This step is accomplished by knowing how to read a map; recognize and identify specific terrain and other features; determine and estimate direction; pace, measure, and estimate distances; and both plot and estimate a position by resection.

    b. **Plan the Route (Step 2).** Depending upon the size of the unit and the length and type of movement to be conducted, several factors should be considered in selecting a good route or routes to be followed. These include—

- Travel time.
- Travel distance.
- Maneuver room needed.
- Trafficability.
- Load-bearing capacities of the soil.
- Energy expenditure by troops.
- The factors of METT-TC.
- Tactical aspects of terrain (OCOKA).
- Ease of logistical support.
- Potential for surprising the enemy.

- Availability of control and coordination features.
- Availability of good checkpoints and steering marks.

In other words, the route must be the result of careful map study and should address the requirements of the mission, tactical situation, and time available. It must also provide for ease of movement and navigation.

(1) Three route-selection criteria that are important for small-unit movements are cover, concealment, and the availability of reliable checkpoint features. The latter is weighted even more heavily when selecting the route for a night operation. The degree of visibility and ease of recognition (visual effect) are the key to the proper selection of these features.

(2) The best checkpoints are linear features that cross the route. Examples include perennial streams, hard-top roads, ridges, valleys, railroads, and power transmission lines. Next, it is best to select features that represent elevation changes of at least two contour intervals such as hills, depressions, spurs, and draws. Primary reliance upon cultural features and vegetation is cautioned against because they are most likely to have changed since the map was last revised.

(3) Checkpoints located at places where changes in direction are made mark your *decision points*. Be especially alert to see and recognize these features during movement. During preparation and planning, it is especially important to review the route and anticipate where mistakes are most likely to be made so they can be avoided.

(4) Following a valley floor or proceeding near (not on) the crest of a ridgeline generally offers easy movement, good navigation checkpoints, and sufficient cover and concealment. It is best to follow terrain features whenever you can—not to fight them.

(5) A lost or a late arriving unit, or a tired unit that is tasked with an unnecessarily difficult move, does not contribute to the accomplishment of a mission. On the other hand, the unit that moves too quickly and carelessly into a destructive ambush or leaves itself open to air strikes also has little effect. Careful planning and study are required each time a movement route is to be selected.

c. **Stay on the Route (Step 3).** In order to know that you are still on the correct route, you must be able to compare the evidence you encounter as you move according to the plan you developed on the map when you selected your route. This may include watching your compass reading (dead reckoning) or recognizing various checkpoints or landmarks from the map in their anticipated positions and sequences as you pass them (terrain association). A better way is to use a combination of both.

d. **Recognize the Objective (Step 4).** The destination is rarely a highly recognizable feature such as a dominant hilltop or road junction. Such locations as this are seldom missed by the most inexperienced navigators and are often dangerous places for Soldiers to occupy. The relatively small, obscure places are most likely to be the destinations.

(1) Just how does a Soldier travel over unfamiliar terrain for moderate to great distances and know when he reaches the destination? One minor error, when many are possible, can cause the target to be missed. The answer is simple.

(2) Select a checkpoint (reasonably close to the destination) that is not so difficult to find or recognize. Then plan a short, fine-tuned last leg from the new *expanded objective* to the final destination. For example, you may be able to plan and execute the move as a series of sequenced movements from one checkpoint or landmark to another using both the terrain and a compass to keep you on the correct course. Finally, after arriving at the last checkpoint, you might follow a specific compass azimuth and pace off the relatively short,

known distance to the final, pinpoint destination. This procedure is called *point navigation*. A short movement out from a unit position to an observation post or to a coordination point may also be accomplished in the same manner.

11-6. NAVIGATION METHODS

Staying on the route is accomplished through the use of one or two navigation techniques—dead reckoning and terrain association. These methods are discussed in detail below.

   a. **Moving by Dead Reckoning.** Dead reckoning consists of two fundamental steps. The first is the use of a protractor and graphic scales to determine the direction and distance from one point to another on a map. The second step is the use of a compass and some means of measuring distance to apply this information on the ground. In other words, it begins with the determination of a polar coordinate on a map and ends with the act of finding it on the ground.

   (1) Dead reckoning along a given route is the application of the same process used by a mapmaker as he establishes a measured line of reference upon which to construct the framework of his map. Therefore, triangulation exercises (either resection or intersection) can be easily undertaken by the navigator at any time to either determine or confirm precise locations along or near his route. Between these position-fixes, establish your location by measuring or estimating the distance traveled along the azimuth being followed from the previous known point. You might use pacing, a vehicle odometer, or the application of elapsed time for this purpose, depending upon the situation.

   (2) Most dead reckoned movements do not consist of single straight-line distances because you cannot ignore the tactical and navigational aspects of the terrain, enemy situation, natural and man-made obstacles, time, and safety factors. Another reason most dead reckoning movements are not single straight-line distances is because compasses and pace counts are imprecise measures. Error from them compounds over distance; therefore, you could soon be far from your intended route even if you performed the procedures correctly. The only way to counteract this phenomenon is to reconfirm your location by terrain association or resection. Routes planned for dead reckoning generally consist of a series of straight-line distances between several checkpoints with perhaps some travel running on or parallel to roads or trails.

   (3) There are two advantages to dead reckoning. First, dead reckoning is easy to teach and to learn. Second, it can be a highly accurate way of moving from one point to another if done carefully over short distances, even where few external cues are present to guide the movements.

   (4) During daylight, across open country, along a specified magnetic azimuth, never walk with the compass in the open position and in front of you. Because the compass will not stay steady or level, it does not give an accurate reading when held or used this way. Begin at the start point and face with the compass in the proper direction, then sight in on a landmark that is located on the correct azimuth to be followed. Close the compass and proceed to that landmark. Repeat the process as many times as necessary to complete the straight-line segment of the route.

   (5) The landmarks selected for this purpose are called *steering marks*, and their selection is crucial to success in dead reckoning. Steering marks should never be determined from a map study. They are selected as the march progresses and are commonly on or near the highest points visible along the azimuth line you are following when they are selected. They

may be uniquely shaped trees, rocks, hilltops, posts, towers, and buildings—anything that can be easily identified. If you do not see a good steering mark to the front, you might use a back azimuth to some feature behind you until a good steering mark appears out in front. Characteristics of a good steering mark are:

(a) It must have some characteristics about it, such as color, shade of color, size, or shape (preferably all four), that will assure you that it will continue to be recognized as you approach it.

(b) If several easily distinguished objects appear along your line of march, the best steering mark is the most distant object. This procedure enables you to travel farther with fewer references to the compass. If you have many options, select the highest object. A higher mark is not as easily lost to sight as is a lower mark that blends into the background as you approach it. A steering mark should be continuously visible as you move toward it.

(c) Steering marks selected at night must have even more unique shapes than those selected during daylight. As darkness approaches, colors disappear and objects appear as black or gray silhouettes. Instead of seeing shapes, you begin to see only the general outlines that may appear to change as you move and see the objects from slightly different angles.

(6) Dead reckoning without natural steering marks is used when the area through which you are traveling is devoid of features, or when visibility is poor. At night, it may be necessary to send a member of the unit out in front of your position to create your own steering mark in order to proceed. His position should be as far out as possible to reduce the number of chances for error as you move. Arm-and-hand signals or a radio may be used in placing him on the correct azimuth. After he has been properly located, move forward to his position and repeat the process until some steering marks can be identified or until you reach your objective.

(7) When handling obstacles/detours on the route, follow these guidelines:

(a) When an obstacle forces you to leave your original line of march and take up a parallel one, always return to the original line as soon as the terrain or situation permits.

(b) To turn clockwise (right) 90 degrees, you must add 90 degrees to your original azimuth. To turn counterclockwise (left) 90 degrees from your current direction, you must subtract 90 degrees from your present azimuth.

(c) When making a detour, be certain that only paces taken toward the final destination are counted as part of your forward progress. They should not be confused with the local pacing that takes place perpendicular to the route in order to avoid the problem area and in returning to the original line of march after the obstacle has been passed.

(8) Sometimes a steering mark on your azimuth of travel can be seen across a swamp or some other obstacle which you can simply walk around. Dead reckoning can then begin at that point. If there is no obvious steering mark to be seen across the obstacle, perhaps one can be located to the rear. Compute a back azimuth to this point and later sight back to it once the obstacle has been passed in order to get back on track.

(9) You can use the deliberate offset technique. Highly accurate distance estimates and precision compass work may not be required if the destination or an intermediate checkpoint is located on or near a large linear feature that runs nearly perpendicular to your direction of travel. Examples include roads or highways, railroads, power transmission lines, ridges, or streams. In these cases, you should apply a deliberate error (offset) of about 10 degrees to the azimuth you planned to follow and then move, using the lensatic compass as a guide, in that direction until you encounter the linear feature. You will know exactly which way to turn

(left or right) to find your destination or checkpoint, depending upon which way you planned your deliberate offset.

(10) Because no one can move along a given azimuth with absolute precision, it is better to plan a few extra steps than to begin an aimless search for the objective once you reach the linear feature. If you introduce your own mistake, you will certainly know how to correct it. This method will also cope with minor compass errors and the slight variations that always occur in the earth's magnetic field.

(11) There are disadvantages to dead reckoning. The farther you travel by dead reckoning without confirming your position in relation to the terrain and other features, the more errors you will accumulate in your movements. Therefore, you should confirm and correct your estimated position whenever you encounter a known feature on the ground that is also on the map. Periodically, you should accomplish a resection triangulation using two or more known points to pinpoint and correct your position on the map. Pace counts or any type of distance measurement should begin anew each time your position is confirmed on the map.

(a) It is dangerous to select a single steering mark, such as a distant mountaintop, and then move blindly toward it. What will you do if you must suddenly call for fire support or a medical evacuation? You must periodically use resection and terrain association techniques to pinpoint your location along the way.

(b) Steering marks can be farther apart in open country, thereby making navigation more accurate. In areas of dense vegetation, however, where there is little relief, during darkness, or in fog, your steering marks must be close together. This, of course, introduces more chance for error.

(c) Finally, dead reckoning is time-consuming and demands constant attention to the compass. Errors accumulate easily and quickly. Every fold in the ground and detours as small as a single tree or boulder also complicate the measurement of distance.

b. **Moving by Terrain Association**. The technique of moving by terrain association is more forgiving of mistakes and far less time-consuming than dead reckoning. It best suits those situations that call for movement from one area to another. Once an error has been made in dead reckoning, you are off the track. Errors made using terrain association are easily corrected, however, because you are comparing what you expected to see from the map to what you do see on the ground. Errors are anticipated and will not go unchecked. You can easily make adjustments based upon what you encounter. After all, you do not find the neighborhood grocery store by dead reckoning—you adjust your movements according to the familiar landmarks you encounter along the way (Figure 11-8). Periodic position-fixing through either plotted or estimated resection will also make it possible to correct your movements, call for fire, or call in the locations of enemy targets or any other information of tactical or logistical importance.

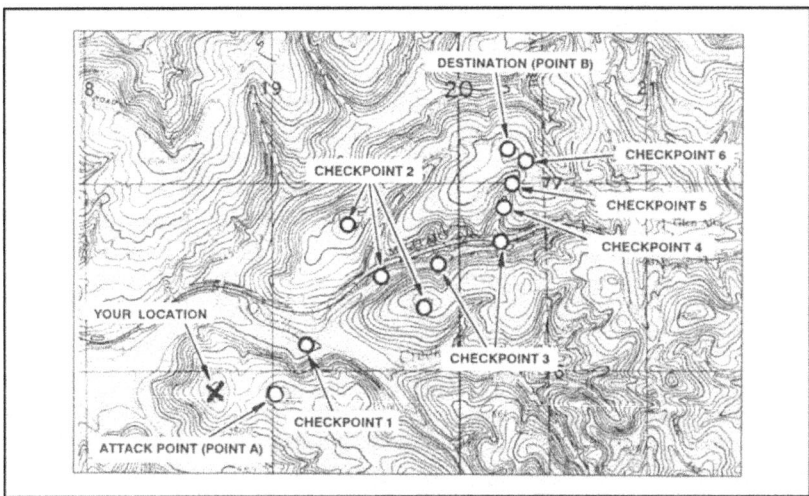

Figure 11-8. Terrain association navigation.

(1) *Identifying and Locating Selected Features.* Being able to identify and locate the selected features, both on the map and on the ground, are essential to the success in moving by terrain association. The following rules may prove helpful.

(a) Be certain the map is properly oriented when moving along the route and use the terrain and other features as guides. The orientation of the map must match the terrain or it can cause confusion.

(b) To locate and identify features being used to guide the movement, look for the steepness and shape of the slopes, the relative elevations of the various features, and the directional orientations in relation to your position and to the position of the other features you can see.

(c) Make use of the additional cues provided by hydrography, culture, and vegetation. All the information you can gather will assist you in making the move. The ultimate test and the best practice for this movement technique is to go out in the field and use it. The use of terrain, other natural features, and any man-made objects that appear both on the map and on the ground must be practiced at every opportunity. There is no other way to learn or retain this skill.

(2) *Using Handrails, Catching Features, and Navigational Attack Points.* First, because it is difficult to dead reckon without error over long distances with your compass, the alert navigator can often gain assistance from the terrain.

(a) *Handrails* are linear features like roads or highways, railroads, power transmission lines, ridgelines, or streams that run roughly parallel to your direction of travel. Instead of using precision compass work, you can rough compass without the use of steering marks for as long as the feature travels with you on your right or left. It acts as a handrail to guide the way.

(b) Second, when you reach the point where either your route or the handrail changes direction, you must be aware that it is time to go your separate ways. Some prominent feature located near this point is selected to provide this warning. This is called a *catching feature*; it can also be used to tell you when you have gone too far.

(c) Third, the catching feature may also be your *navigational attack point*; this point is the place where area navigation ends and point navigation begins. From this last easily identified checkpoint, the navigator moves cautiously and precisely along a given azimuth for a specified distance to locate the final objective. The selection of this navigational attack point is important. A distance of 500 meters or less is most desirable.

(3) **Recognizing the Disadvantages of Terrain Association.** The major disadvantage to navigation by terrain association is that you must be able to interpret the map and analyze the world around you. Recognition of terrain and other features, the ability to determine and estimate direction and distance, and knowing how to do quick in-the-head position fixing are skills that are more difficult to teach, learn, and retain than those required for dead reckoning.

c. **Combination of Techniques.** Actually, the most successful navigation is obtained by combining the techniques described above. Constant orientation of the map and continuous observation of the terrain in conjunction with compass-read azimuths, and distance traveled on the ground compared with map distance, used together make reaching a destination more certain. One should not depend entirely on compass navigation or map navigation; either or both could be lost or destroyed.

NOTE: See Appendix E for information on orienteering.

11-7. NIGHT NAVIGATION

Darkness presents its own characteristics for land navigation because of limited or no visibility. However, the techniques and principles are the same as those used for day navigation. The success in nighttime land navigation depends on rehearsals during the planning phase before the movement, such as detailed analysis of the map to determine the type of terrain in which the navigation is going to take place, and the predetermination of azimuths and distances. Night vision devices (Appendix G) can greatly enhance night navigation.

a. The basic technique used for nighttime land navigation is dead reckoning with several compasses recommended. The point man is in front of the navigator but just a few steps away for easy control of the azimuth. Smaller steps are taken during night navigation, so remember, the pace count is different. It is recommended that a pace count obtained by using a predetermined 100-meter pace course be used at night.

b. Navigation using the stars is recommended in some areas; however, a thorough knowledge of constellations and location of stars is needed (Chapter 9, paragraph 9-5c). The four cardinal directions can also be obtained at night using the same technique described for the shadow-tip method—just use the moon instead of the sun. In this case, the moon must be bright enough to cast a shadow.

CHAPTER 12
MOUNTED LAND NAVIGATION

A vehicle commander should be able to navigate from one point on the ground to another with or without a compass. If separated from his unit and given an azimuth and distance from their position to his, he should be able to reach the unit and continue the mission. To move effectively while mounted, he must know the principles of mounted navigation.

12-1. PRINCIPLES
The principles of land navigation while mounted are basically the same as while dismounted. The major difference is the speed of travel. Walking between two points may take one hour, but riding the same distance may only take 15 minutes. To be effective at mounted land navigation, the travel speed must be considered.

12-2. NAVIGATOR'S DUTIES
The duties of a navigator are so important and exacting that he should not be given any other duties. The leader should never try to be the navigator, since his normal responsibilities are heavy, and one or the other job would suffer.

    a. **Assembling Equipment.** Before the mission starts, the navigator must gather all the equipment that will help him perform his job (maps, pencils, and so forth).

    b. **Servicing Equipment.** The navigator is responsible for making sure that all the equipment he may use or require is working.

    c. **Recording Data for Precise Locations.** During movement, the navigator must make sure that the correct direction and distance are recorded and followed. Grid coordinates of locations must be recorded and plotted.

    d. **Supplying Data to Subordinate Leaders.** During movement, any change in direction or distance must be given to the subordinate leaders in sufficient time to allow them to react.

    e. **Maintaining Liaison with the Commander.** The commander normally selects the route that he wants to use. The navigator is responsible for following that route; however, there may be times when the route must be changed during a tactical operation. For this reason, the navigator must maintain constant communication with the commander. The navigator must inform the commander when checkpoints are reached, when a change in direction of movement is required, and how much distance is traveled.

12-3. MOVEMENT
When preparing to move, the effects of terrain on navigating mounted vehicles must be determined. You will cover great distances very quickly, and you must develop the ability to estimate the distance you have traveled. Remember that 0.1 mile is roughly 160 meters, and 1 mile is about 1,600 meters or 1.6 kilometers. Having a mobility advantage helps while navigating. If you get disoriented, mobility makes it much easier to move to a point where you can reorient yourself.

NOTE: To convert kilometers per hour to miles per hour, multiply by 0.62 (for example, 9 kilometers per hour x 0.62 = 5.58 miles per hour). To convert miles per hour to kilometers per hour, divide miles per hour by 0.62 (for example, 10 miles per hour ÷ 0.62 = 16.12 kilometers per hour).

a. **Consider Vehicle Capabilities.** When determining a route to be used when mounted, consider the capabilities of the vehicles to be used. Most military vehicles are limited in the degree of slope they can climb and the type of terrain they can negotiate. Swamps, thickly wooded areas, or deep streams may present no problems to dismounted Soldiers, but the same terrain may completely stop mounted Soldiers. The navigator must consider this when selecting a route.

(1) Most vehicles can knock down a tree. The bigger the vehicle, the bigger the tree it can knock down. Vehicles cannot knock down several trees at once. It is best to find paths between trees that are wide enough for your vehicle. Military vehicles are designed to climb 60-percent slopes on a dry, firm surface (Figure 12-1).

(2) You can easily determine approximate slope by looking at the route you have selected on a map. A contour line in any 100 meters of map distance on that route indicates a 10-percent slope, two contour lines indicate 20–percent slope, and so forth. If there are four contour lines in any 100 meters, look for another route.

(3) Side slope is even more important than the slope you can climb. Normally, a 30-percent slope is the maximum in good weather. When traversing a side slope, progress slowly and without turns. Rocks, stumps, or sharp turns can cause you to throw the downhill track under the vehicle, which would mean a big recovery task.

(4) For tactical reasons, you will often want to move in draws or valleys because they provide cover. However, side slopes force you to move slowly.

NOTE: The above figures are true for a 10-meter or a 20-foot contour interval. If the map has a different contour interval, just adjust the arithmetic. For instance, with one contour line in 100 meters, a 20-meter interval would give a 20-percent slope.

b. **Know the Effects of Weather on Vehicle Movement.** Weather can halt mounted movement. Snow and ice are obvious dangers, but more significant is the effect of rain and snow on the load-bearing ability of soil. Cross-country vehicles may be restricted to road movement in heavy rain. If it has rained recently, adjust your route to avoid flooded or muddy areas. A mired vehicle only hinders combat capability.

c. **Prepare Before Movement.** Locate the start point and finish point on the map. Determine the map's grid azimuth from start point to finish point and convert it to a magnetic azimuth. Determine the distance between the start point and finish point or any intermediate points on the map and make a thorough map reconnaissance of that area.

Figure 12-1. Tracked vehicle capabilities.

12-4. TERRAIN ASSOCIATION NAVIGATION

Terrain association is currently the most widely used method of navigation. The navigator plans his route so that he moves from terrain feature to terrain feature. An automobile driver in a city uses this technique as he moves along a street or series of streets, guiding on intersections or features such as stores and parks. Like the driver, the navigator selects routes or *streets* between key points or *intersections*. These routes must be capable of sustaining the travel of the vehicle or vehicles, should be relatively direct, and should be easy to follow. In a typical move, the navigator determines his location, determines the location of his objective, notes the position of both on his map, and then selects a route between the two. After examining the terrain, he adjusts the route using the following actions:

    a. **Consider Tactical Aspects.** Avoid skylining, select key terrain for overwatch positions, and select concealed routes.

    b. **Consider Ease of Movement.** Use the easiest possible route and bypass difficult terrain. Remember that a difficult route is harder to follow, is noisier, causes more wear and tear (and possible recovery problems), and takes more time. Tactical surprise is achieved by doing the unexpected. Try to select an axis or corridor instead of a specific route. Make sure you have enough maneuver room for the vehicles (Figure 12-2).

FM 3-25.26

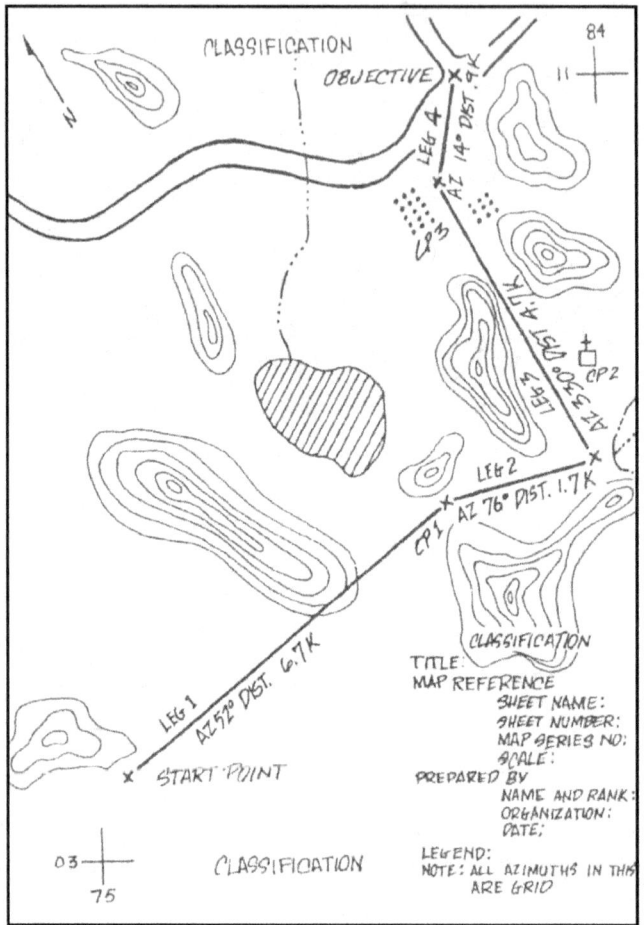

Figure 12-2. Primary route.

c. **Use Terrain Features as Checkpoints.** These checkpoints must be easily recognizable in the light and weather conditions and at the speed at which you will move. You should be able to find a terrain feature from your location that can be recognized from almost anywhere and used as a guide. (For example, checkpoint 2, the church and checkpoint 3, the orchard in Figure 12-2.)

(1) The best checkpoints are linear features that cross your route. Use streams, rivers, hard-top roads, ridges, valleys, and railroads.

(2) The next best checkpoints are elevation changes such as hills, depressions, spurs, and draws. Look for two contour lines of change. You will not be able to spot less than two lines of change while mounted.

(3) In wooded terrain, try to locate checkpoints at no more than 1,000-meter intervals. In open terrain, you may go to about 5,000 meters.

d. **Follow Terrain Features.** Movement and navigation along a valley floor or near (not necessarily on) the crest of a ridgeline is easiest.

e. **Determine Directions.** Break the route down into smaller segments and determine the rough directions that will be followed. You do not need to use the compass; just use the main points of direction (north, northeast, east, and so forth). Before moving, note the location of the sun and locate north. Locate changes of direction, if any, at the checkpoints picked.

f. **Determine Distance.** Determine the total distance to be traveled and the approximate distance between checkpoints. Plan to use the vehicle odometer to keep track of distance traveled. Use the pace-count method and keep a record of the distance traveled. When using a pace count, convert from map distance to ground distance by adding the conversion factors of 20 percent for cross-country movement.

g. **Make Notes.** Mental notes are usually adequate. Try to imagine what the route is like and remember it.

h. **Plan to Avoid Errors.** Restudy the route selected. Try to determine where errors are most apt to occur and how to avoid any trouble.

i. **Use a Logbook.** When the routes have been selected and the navigator has divided the distance to be traveled into legs, prepare a logbook. The logbook is an informal record of the distance and azimuth of each leg, with notes to aid the navigator in following the correct route. The notes list easily identifiable terrain features at or near the point where the direction of movement changes (Figure 12-3).

ODOMETER READING AT START	ODOMETER READING AT FINISH	DISTANCE IN MILES	AZIMUTH	DEVIATION CORRECTION	NOTES

Figure 12-3. Sample of a logbook format.

12-5. DEAD RECKONING NAVIGATION

Dead reckoning is moving a set distance along a set line. Generally, it involves moving so many meters along a set line, usually an azimuth in degrees. There is no accurate method of determining a direction in a moving vehicle. A magnetic vehicle-heading reference unit may be available in a few years; for now, use a compass.

a. **With Steering Marks.** This procedure is the same for vehicle travel as on foot.

(1) The navigator dismounts from the vehicle and moves away from the vehicle (at least 18 meters).

(2) He sets the azimuth on the compass and picks a steering mark (rock, tree, hilltop) in the direction on that azimuth (Figure 12-4).

(3) He remounts and has the driver identify the steering mark and proceed to it in as straight a line as possible.

(4) On arrival at the steering mark or on any changes in direction, the navigator repeats the first three steps for the next leg of travel.

Figure 12-4. Determining an azimuth, dismounted.

b. **Without Steering Marks.** This procedure is used only on flat, featureless terrain.

(1) The navigator dismounts from the vehicle, which is oriented in the direction of travel, and moves at least 18 meters to the front of the vehicle.

(2) He faces the vehicle and reads the azimuth to the vehicle. By adding or subtracting 180 degrees, he determines the forward azimuth (direction of travel).

(3) On order from the navigator, the driver drives on a straight line to the navigator.

(4) The navigator remounts the vehicle, holds the compass as it will be held while the vehicle is moving, and reads the azimuth in the direction of travel.

(5) The compass will swing off the azimuth determined and pick up a constant deviation. For example, the azimuth was 75 degrees while you were away from the vehicle. When you remounted and your driver drove straight forward, your compass showed 67 degrees. You have a deviation of -8 degrees. All you need to do is maintain that 67-degree compass heading to travel on a 75-degree magnetic heading.

(6) At night, the same technique can be used. From the map, determine the azimuth you are to travel. Convert the grid azimuth to a magnetic azimuth. Line the vehicle up on that azimuth, then move well in front of it. Be sure it is aligned correctly. Then mount, have the driver move slowly forward, and note the deviation. If the vehicle has a turret, the above procedure works unless you traverse the turret; this changes the deviation.

(7) The distance factor in dead reckoning is easy. Just determine the map distance to travel and add 20 percent to convert to ground distance. Use your vehicle odometer to be sure you travel the proper distance.

12-6. STABILIZED TURRET ALIGNMENT NAVIGATION

Another method, if you have a vehicle with a stabilized turret, is to align the turret on the azimuth you wish to travel, then switch the turret stabilization system on. The gun tube remains pointed at your destination no matter which way you turn the vehicle. This technique has been proven; it works. It is not harmful to the stabilization system. It is subject to stabilization drift, so use it for no more than 5,000 meters before resetting.

NOTE: If you have to take the turret off-line to engage a target, you will have to re-do the entire process.

12-7. COMBINATION NAVIGATION

Some mounted situations may call for you to combine and use both methods. Just remember the characteristics of each.

a. Terrain association is fast, is error-tolerant, and is best under most circumstances. It can be used day or night if you are proficient in it.

b. Dead reckoning is very accurate if you do everything correctly. You must be very precise. It is also slow, but it works on very flat terrain.

c. You frequently will combine both. You may use dead reckoning to travel across a large, flat area to a ridge, then use terrain association for the rest of the move.

d. You must be able to use both methods. You should remember that your probable errors, in order of frequency, will be—
- Failure to determine distances to be traveled.
- Failure to travel the proper distance.
- Failure to properly plot or locate the objective.
- Failure to select easily recognized checkpoints or landmarks.
- Failure to consider the ease of movement factor.

This Page intentionally left blank.

CHAPTER 13
NAVIGATION IN DIFFERENT TYPES OF TERRAIN

The information, concepts, and skills already presented will help you to navigate anywhere in the world; however, there are some special considerations and helpful hints that may assist you in various special environments. The following information is not doctrine.

13-1. DESERT TERRAIN

About 5 percent of the earth's land surface is covered by deserts (Figure 13-1). Deserts are large arid areas with little or no rainfall during the year. There are three types of deserts—mountain, rocky plateau, and sandy or dune deserts. All types of forces can be deployed in the desert. Armor and mechanized infantry forces are especially suitable to desert combat except in rough mountainous terrain where light infantry may be required. Airborne, air assault, and motorized forces can also be advantageously employed to exploit the vast distances characteristic of desert warfare.

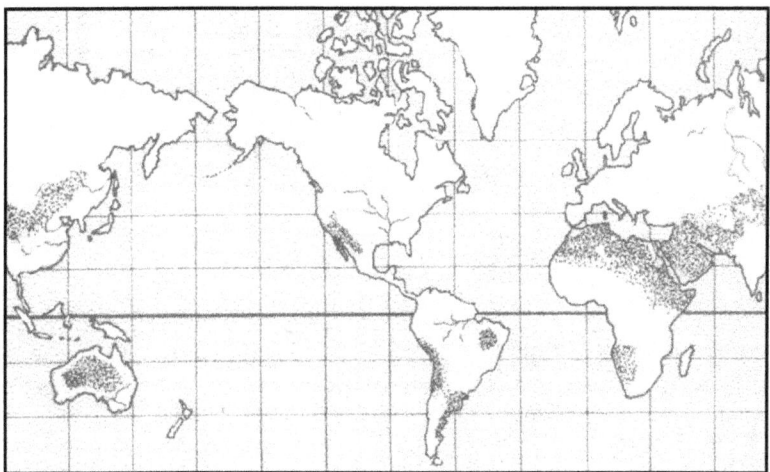

Figure 13-1. Deserts.

a. **Desert Regions.** In desert regions, terrain varies from nearly flat to lava beds and salt marshes. Mountain deserts contain scattered ranges or areas of barren hills or mountains. Table 13-1 (page 13-2) lists some of the world's major desert regions and their locations.

DESERT REGION	LOCATION
Sahara	North Africa
Kalahari	Southwest Africa
Arabian	Southwest Asia
Gobi	Mongolia and Northern China
Rub'al Khali	South Arabia
Great Basin, Colorado, Chichuahua, Yuma Sonoran, and Mohave	Northern Mexico and Western United States
Takla Makan	Northern China
Kyzyl Kum	Southwest USSR
Kara Kum	Southwest USSR
Syrian	Saudi Arabia, Jordan, and Iraq
Great Victoria	Western and South Australia
Great Sandy	Northwestern Australia
Patagonia	Southern Argentina and Chile
Atacama	Northern Chile

Table 13-1. Location of major desert regions.

(1) Finding the way in a desert presents some degree of difficulty for a person who has never been exposed to this environment. Desert navigators have learned through generations of experience.

(2) Normally, desert people are nomadic, constantly moving in caravans. Navigating becomes second nature to them. Temperature in the tropical deserts reaches an average of 110 to 115 degrees during the day, so most navigation takes place at night using the stars. Most deserts have some prevailing winds during the seasons. Such winds arrange the sand dunes in a specific pattern that allows the navigator to determine the four cardinal directions. He may also use the sun's shadow-tip method.

(3) A sense of direction can be obtained by watching desert animals on their way to and from water holes (oases). Water, navigation, and survival are closely related in desert areas. Most deserts have pigeons or doves, and their drinking habits are important to the navigator. As a rule, these birds never drink in the morning or during the day, making their evening flights the most important. When returning from the oases, their bodies are heavier from drinking and their flight is accompanied by a louder flapping of their wings.

(4) Visibility is also an important factor in the desert, especially in judging distance. The absence of trees or other features prevents comparison between the horizon and the skyline.

b. **Interpretation and Analysis.** Many desert maps are inaccurate, which makes up-to-date air, aerial photo, and ground reconnaissance necessary. In desert mountain areas contour intervals are generally large, so many of the intermediate relief features are not shown.

(1) The desert normally permits observation and fire to maximum ranges. The terrain is generally wide open and the exceptionally clear atmosphere offers excellent long-range visibility. Combine this with a powerful sun and low cloud density and you have nearly unlimited light and visual clarity, which often contribute to gross underestimations of ranges. Errors of up to 200 or 300 percent are not uncommon. However, visibility conditions may be severely affected by sandstorms and mirages (heat shimmer caused by air rising from the extremely hot daytime desert surface), especially if the observer is looking into the sun through magnifying optical instruments.

(2) Cover can be provided only by terrain feature masking because of the lack of heavy vegetation and man-made objects. It only takes a few meters of relief to provide cover. Concealment in the desert is related to the following factors:

(a) *Shape.* To prevent observation by the enemy, try to alter the standard shapes of vehicles so they and their shadows are not instantly recognizable.

(b) *Shine.* Shine or glitter is often the first thing that attracts the observer's eye to movement many kilometers away. It must be eliminated.

(c) *Color and Texture.* All equipment should either be pattern painted or mudded to blend in with the terrain.

(d) *Light and Noise.* Light and noise discipline are essential because sound and light travel great distances in the desert.

(e) *Heat.* Modern heat image technology makes shielding heat sources an important consideration when trying to hide from the enemy. This technology is especially important during night stops.

(f) *Movement.* Movement itself creates a great deal of noise and dust, but a rapid execution using all the advantages the topography offers can help conceal it.

c. **Navigation.** When operating in the broad basins between mountain ranges or on rocky plateau deserts, there are frequently many terrain features to guide your movement. Observing these known features over great distances may provide a false sense of security in determining your precise location unless you frequently confirm your location by resection or referencing close-in terrain features. It is not uncommon to develop errors of several kilometers when casually estimating a position in this manner. Obviously, this can create many problems when attempting to locate a small checkpoint or objective, calling for combat support (CS), reporting operational or intelligence information, or meeting combat service support (CSS) requirements.

(1) When operating in an area with few visual cues, such as a sandy or dune desert, or when visibility is restricted by a sandstorm or darkness, you must proceed by dead reckoning. The four steps and two techniques for navigation presented earlier remain valid in the desert. However, understanding the desert's special conditions is extremely helpful as you apply the techniques.

(2) Tactical mobility and speed are key to successful desert operations. Obstacles and areas such as lava beds or salt marshes, which preclude surface movements, do exist. But most deserts permit two-dimensional movement by ground forces similar to that of a naval task force at sea. Speed of execution is essential. Everyone moves farther and faster on the desert. Special navigation aids sometimes used in the desert include:

(a) *Sun Compass.* The sun compass can be used on moving vehicles and sextants. It requires accurate timekeeping. However, the deviation on a magnetic compass that is caused by the metal and electronics in the vehicle is usually less than +10 degrees.

(b) *Gyro Compass.* The gun azimuth stabilizer is in fact a gyro compass. If used on fairly flat ground, it is useful for maintaining direction over limited distances.

(c) *Fires.* Planned tracer fire or mortar and artillery concentrations (preferably smoke during the day and illumination at night) provide useful checks on estimated locations.

(d) *Pre-positioned Lights.* This method consists of placing two or more searchlights far apart, behind the line of contact, beyond enemy artillery range, and concealed from enemy ground observation. Units in the area can determine their own locations through resection, using the vertical beams of the lights. These lights must be moved on a time schedule known to all friendly units.

(3) The sand, hard-baked ground, rocky surfaces, thorny vegetation, and heat generally found in the desert impose far greater demands for maintenance than you would plan for in temperate regions. It may also take longer to perform that maintenance.

13-2. MOUNTAIN TERRAIN

Mountains are generally understood to be larger than hills. Rarely do mountains occur individually; in most cases, they are found in elongated ranges or circular groups. When they are linked together, they constitute a mountain system (Figure 13-2). Light forces (infantry, airborne, and air assault forces) can operate effectively in mountainous regions because they are not limited by terrain. Heavy forces must operate in passes and valleys that are negotiable by vehicle.

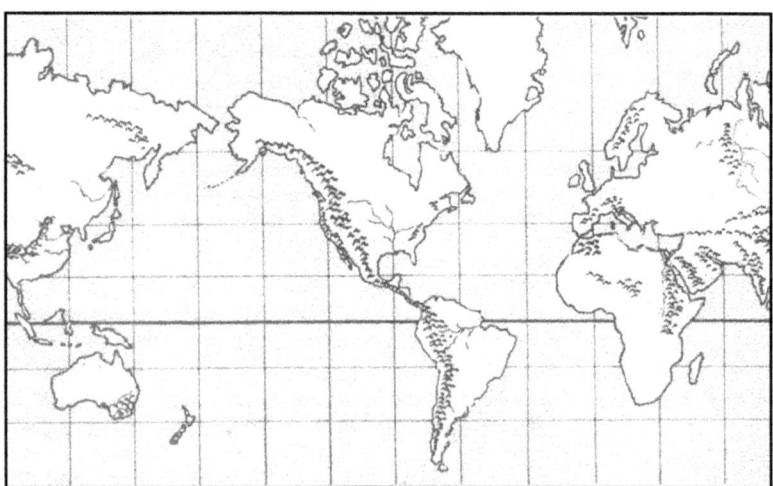

Figure 13-2. Mountain systems.

a. **Characteristics.** Mountain systems are characterized by high, inaccessible peaks and steep slopes. Depending on the altitude, they may be snow covered. Prominent ridges and large valleys are also found. Navigating in this type of terrain is not difficult providing you make a careful examination of the map and the terrain.

b. **Major Mountain Systems.** Table 13-2 lists the major mountain systems and their locations.

MOUNTAIN SYSTEM	LOCATION
Andes	Central and South America
Rockies	North America (USA and Canada)
Appalachians	North America (USA and Canada)
Alps	Central Europe
Himalayas	Asia
Caucasus	Western Asia and Europe (Russia)

Table 13-2. Location of major mountain systems.

c. **Minor Systems.** Some minor mountain systems are located in Antarctica, Hawaii, Japan, New Zealand, and Oceania.

d. **Climate.** Because of the elevations, it is always colder (3 to 5 degrees per 300-meter gain in altitude) and wetter than you might expect. Wind speeds can increase the effects of the cold even more. Sudden severe storms and fog are encountered regularly. Below the tree line, vegetation is heavy because of the extra rainfall and the fact that the land is rarely cleared for farming.

e. **Interpretation and Analysis.** The heights of mountainous terrain permit excellent long-range observation. However, rapidly fluctuating weather with frequent periods of high winds, rain, snow, or fog may limit visibility. Also, the rugged nature of the terrain frequently produces significant dead space at mid ranges.

(1) Reduced mobility, compartmented terrain, and the effects of rapidly changing weather increase the importance of air, ground, aerial photo, and map reconnaissance. Since mountain maps often use large contour intervals, microrelief interpretation and detailed terrain analysis require special emphasis.

(2) At first glance, some mountainous terrain may not appear to offer adequate cover and concealment; however, you can improve the situation. When moving, use rock outcroppings, boulders, and heavy vegetation for cover and concealment; use terrain features to mask maneuvers. Use harsh weather, which often obscures observation, to enhance concealment.

(3) Since there are only a few routing options, all-round security must be of primary concern. Natural obstacles are everywhere, and the enemy can easily construct more.

f. **Navigation.** Existing roads and trails offer the best routes for movement. Off-road movement may enhance security provided there is detailed reconnaissance, photo intelligence, or information from local inhabitants to ensure the route is negotiable. Again, the four steps and two techniques for navigation presented earlier remain valid in the mountains. Nevertheless, understanding the special conditions and the terrain will help you navigate. Other techniques that are sometimes helpful in mountains are:

FM 3-25.26

(1) *Aspect of Slope*. To determine the aspect of slope, take a compass reading along an imaginary line that runs straight down the slope. It should cut through each of the contour lines at about a 90-degree angle. By checking the map and knowing the direction of slope where you are located, you will be able to keep track of your location, and it will help guide your cross-country movement even when visibility is poor.

(2) *Use of an Altimeter*. Employment of an altimeter with calibrations on the scale down to 10 or 20 meters is helpful to land navigators moving in areas where radical changes in elevation exist. An altimeter is a type of barometer that gauges air pressure, except it measures on an adjustable scale marked in feet or meters of elevation rather than in inches or centimeters of mercury. Careful use of the altimeter helps to pinpoint your position on a map through a unique type of resection. Instead of finding your position by using two different directional values, you use one directional value and one elevation value.

13-3. JUNGLE TERRAIN

These large geographic regions are found within the tropics near the equator (Central America, along the Amazon River, South-Eastern Asia and adjacent islands, and vast areas in the middle of Africa and India) (Figure 13-3). Jungles are characterized as rainy, humid areas with heavy layers of tangled, impenetrable vegetation. Jungles contain many species of wildlife (tigers, monkeys, parrots, snakes, alligators, and so forth). The jungle is also a paradise for insects, which are the worst enemy of the navigator because some insects carry diseases (malaria, yellow fever, cholera, and so forth). While navigating in these areas, very little terrain association can be accomplished because of the heavy foliage. Dead reckoning is one of the methods used in these areas. A lost navigator in the jungle can eventually find his way back to civilization by following any body of water with a downstream flow. However, not every civilization found is of a friendly nature.

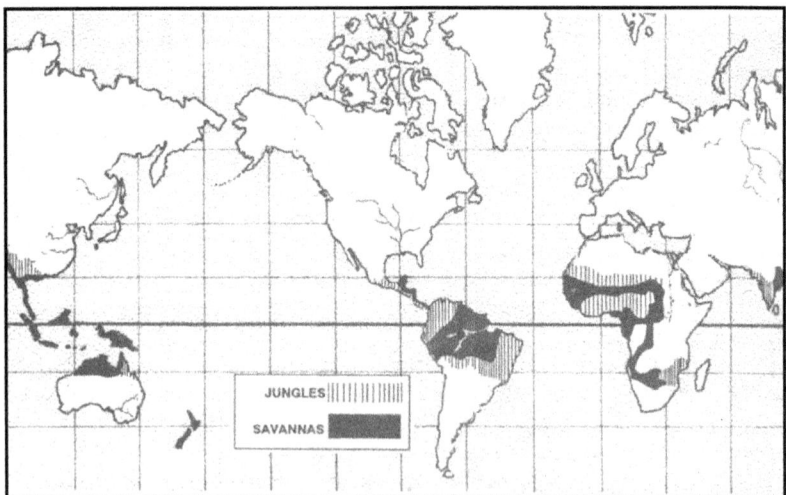

Figure 13-3. Jungles and savannas.

a. **Operations.** Operations in jungles tend to be isolated actions by small forces because of the difficulties encountered in moving and in maintaining contact between units. Divisions can move cross-country slowly; but, aggressive reconnaissance, meticulous intelligence collection, and detailed coordination are required to concentrate forces in this way. More commonly, large forces operate along roads or natural avenues of movement, as was the case in the mountains. Patrolling and other surveillance operations are especially important to ensure security of larger forces in the close terrain of jungles.

(1) Short fields of observation and fire and thick vegetation make maintaining contact with the enemy difficult. The same factors reduce the effectiveness of indirect fire and make jungle combat primarily a fight between infantry forces. Support by air and mechanized forces can be decisive at times, but it will not always be available or effective.

(2) Jungles are characterized by high temperatures, heavy rains, high humidity, and an abundance of vegetation. The climate varies with location. Close to the equator, all seasons are nearly alike with heavy rains all year. Farther from the equator (India and Southeast Asia), there are distinct wet (monsoon) and dry seasons. Both zones have high temperatures (averaging 75 to 95+ degrees Fahrenheit), heavy rainfall (as much as 400+ inches annually), and high humidity (90 percent) all year.

(3) In temperate climates, areas of vegetation are the most likely to be altered and incorrectly portrayed on a map. In jungle areas, the vegetation grows so rapidly that it is more likely to be cleared and make these areas be shown incorrectly.

b. **Interpretation and Analysis.** The jungle environment includes dense forests, grasslands, swamps, and cultivated areas. Forests are classified as primary and secondary based upon the terrain and vegetation. Primary forests include tropical rain forests and deciduous forests. Secondary forests are found at the edges of both rain forests and deciduous forests and in areas where jungles have been cleared and abandoned. These places are typically overgrown with weeds, grasses, thorns, ferns, canes, and shrubs. Movement is especially slow and difficult. The extremely thick vegetation reaches a height of 2 meters and severely limits observation to only a few meters.

(1) Tropical rain forests consist mostly of large trees whose branches spread and lock together to form canopies. These canopies, which can exist at two and three different levels, may form as low as 10 meters from the ground. They prevent direct sunlight from reaching the ground, causing a lack of undergrowth on the jungle floor. Extensive above-ground root systems and hanging vines are common and make vehicular travel difficult; foot movement is easier. Ground observation is limited to about 50 meters and air observation is nearly impossible.

(2) Deciduous forests are in semitropical zones that have both wet and dry seasons. In the wet season, trees are fully leafed; in the dry season, much of the foliage dies. Trees are usually less dense than in rain forests, which allows more sunlight to filter to the ground and produces thick undergrowth. During the wet season, air and ground observation is limited and movement is difficult. During the dry season, both improve.

(3) Swamps are common to all low, jungle areas where there is poor drainage. When navigating in a swampy area, a careful analysis of map and ground should be taken before any movement. The Soldiers should travel in small numbers with only the equipment required for their mission, keeping in mind that they are going to be immersed in water part of the time. The usual technique used in swamp navigation is dead reckoning. There are two basic types of swamps—mangrove and palm. Mangrove swamps are found in coastal areas

wherever tides influence water flow. Mangrove is a shrub-like tree that grows 1 to 5 meters high. These trees have a tangled root system, both above and below the waterline, which restricts movement either by foot or small boat. Observation on the ground and from the air is poor, but concealment is excellent.

(4) Grassy plains or savannas are generally located away from the equator but within the tropics. These vast land areas are characterized by flatlands with a different type of vegetation than jungles. They consist mainly of grasses (ranging from 1 to more than 12 feet in height), shrubs, and isolated trees. The most difficult areas to navigate are the ones surrounded by tall grass (elephant grass); however, vehicles can negotiate here better than in some areas. There are few or no natural features to navigate by, making dead reckoning or navigation by stars the only technique for movement. Depending on the height of the grass, ground observation may vary from poor to good. Concealment from air observation is poor for both Soldiers and vehicles.

(5) *Bamboo* stands are common throughout the tropics. They should be bypassed whenever possible. They are formidable obstacles for vehicles, and Soldier movement through them is slow, exhausting, and noisy.

(6) Cultivated areas exist in jungles also. They range from large, well-planned, well-managed farms and plantations to small tracts cultivated by farmers. The three general types of cultivated areas are rice paddies, plantations, and small farms.

c. **Navigation.** Areas such as jungles are generally not accurately mapped because heavy vegetation makes aerial surveys difficult. The ability to observe terrain features, near or far, is extremely limited. The navigator must rely heavily upon his compass and the dead reckoning technique when moving in the jungle. Navigation is further complicated by the inability to make straight-line movements. Terrain analysis, constant use of the compass, and an accurate pace count are essential to navigation in this environment.

(1) Rates of movement and pace counts are particularly important to jungle navigators. The most common error is to overestimate the distance traveled. The distances in Table 13-3 can be used as a rough guide for the maximum distances that might be traveled in various types of terrain during one hour in daylight.

TYPE OF TERRAIN	MAXIMUM DISTANCE (Meters)
Tropical rain forest	up to 1,000
Deciduous forest	500
Secondary jungle	100 to 500
Tall grass	500
Swamps	100 to 500
Rice paddies (wet)	800
Rice paddies (dry)	2,000
Plantations	2,000
Trails	up to 3,000

Table 13-3. Guide for maximum travel distance in jungle environments.

(2) Special navigation strategies that are helpful in jungles include:

(a) *Personal Pace Table.* You should either make a mental or written personal pace table that includes your average pace count per 100 meters for each of the types of terrain through which you are likely to navigate.

(b) *Resection Using Indirect Fire.* Call for mortar or artillery fire (airbursts of white phosphorous or illumination) on two widely separated grids that are not on terrain features like the one you are occupying and are a safe distance from your estimated location. Directions to the airbursts sometimes must be determined by sound.

(c) *Modified Area/Point Navigation.* Even when making primary use of the compass for dead reckoning, you are frequently able to area navigate to an expanded objective, which is easily identified by terrain association. Then, simply develop a short, point-navigation leg to your final destination.

13-4. ARCTIC TERRAIN

Arctic terrain includes those areas that experience extended periods of below freezing temperatures. In these areas, the ground is generally covered with ice or snow during the winter season. Although frozen ground and ice can improve trafficability, a deep accumulation of snow can reduce it. Vehicles and personnel require special equipment and care under these adverse conditions.

a. **Operations.** Both the terrain and the type and size of unit operations vary greatly in arctic areas. In open terrain, armored and mechanized forces will be effective although they will have to plan and train for the special conditions. In broken terrain, forests, and mountains, light forces will predominate as usual. However, foot movement may take up to five times as long as it might under warmer conditions.

b. **Interpretation and Analysis.** Both the terrain and cultural features you may confront in winter may vary to any extreme, as can the weather. The common factor is an extended period of below freezing temperatures. The terrain may be plains, plateaus, hills, or mountains. The climate will be cold, but the weather will vary greatly from place to place. Although most arctic terrain experiences snow, some claim impressive accumulations each season such as the lake-effected snow belts off Lake Ontario near Fort Drum, New York. Other areas have many cold days with sunshine and clear nights, and little snow accumulation.

(1) In areas with distinct local relief and scattered trees or forests, the absence of foliage makes movement by terrain association easier; observation and fields of fire are greatly enhanced except during snowstorms. In relatively flat, open areas covered with snow (especially in bright sunlight), the resulting lack of contrast may interfere with your being able to read the land. With foliage gone, concealment (both from the ground and from the air) is greatly reduced. As in desert areas, you must make better use of the terrain to conceal your movements.

(2) Frozen streams and swamps may no longer be obstacles, and thus identification of avenues of approach may be difficult in winter. However, the concept as to what is key terrain is not likely to be affected.

c. **Navigation.** Special skills may be required in arctic terrain, such as the proper use of winter clothing, skis, and snowshoes; but this does not affect your navigation strategies. There are no special techniques for navigating in arctic terrain. Just be aware of the advantages and disadvantages that may present themselves and make the most of your opportunities while applying the four steps and two techniques for land navigation.

(1) Remember, the highest caliber of leadership is required to ensure that all necessary tasks are performed, that security is maintained, and that Soldiers and their equipment are protected from the physical effects of very low temperatures. There is a great temptation to do less than a thorough job at whatever the task may be when you are very cold.

(2) Night navigation may be particularly enhanced when operating in arctic terrain. Moonlight and starlight on a clear night reflect off the snow, thus enabling you to employ daytime terrain association techniques with little difficulty. Even cloudy winter nights are often brighter than clear moonlit summer nights when the ground is dark and covered with foliage. Movements with complete light discipline (no blackout drives) can often be executed. On the other hand, areas with severe winter climates experience lengthy periods of darkness each day, which may be accompanied by snow and limited visibility.

13-5. URBAN AREAS

The world continues to become more urbanized each year; therefore, it is unlikely that all fighting will be done in rural settings. Major urban areas represent the power and wealth of a particular country in the form of industrial bases, transportation complexes, economic institutions, and political and cultural centers. Therefore, it may be necessary to secure and neutralize them. When navigating in urban places, it is man-made features, such as roads, railroads, bridges, and buildings, that become important while terrain and vegetation become less useful.

   a. **Interpretation and Analysis.** Military operations on urbanized terrain require detailed planning that provides for decentralized execution. As a result of the rapid growth and changes occurring in many urban areas, the military topographic map is likely to be outdated. Supplemental use of commercially-produced city maps may be helpful, or an up-to-date sketch can be made.

   (1) Urbanized terrain normally offers many AAs for mounted maneuver well forward of and leading to urban centers. In the proximity of these built-up areas, however, such approach routes generally become choked by urban sprawl and perhaps by the nature of adjacent natural terrain. Dismounted forces then make the most of available cover by moving through buildings and underground systems, along edges of streets, and over rooftops. Urban areas tend to separate and isolate units, requiring the small-unit leader to take the initiative and demonstrate his skill in order to prevail.

   (2) The urban condition of an area creates many obstacles, and the destruction of many buildings and bridges as combat power is applied during a battle further limits your freedom of movement. Cover and concealment are plentiful, but observation and fields of fire are greatly restricted.

   b. **Navigation.** Navigation in urban areas can be confusing, but there are often many cues that will present themselves as you proceed. They include streets and street signs; building styles and sizes; the urban geography of industrial, warehousing, residential housing, and market districts; man-made transportation features other than streets and roads (rail and trolley lines); and the terrain features and hydrographic features located within the built-up area. Use the following strategies to stay on the route in an urban area.

   (1) *Process Route Descriptions*. Write down or memorize the route through an urban area as a step-by-step process. For example, "Go three blocks north, turn left (west) on a

wide divided boulevard until you go over a river bridge. Turn right (north) along the west bank of the river, and . . . "

(2) *Conceptual Understanding of the Urban Area.* While studying the map and operating in a built-up area, work hard to develop an understanding (mental map) of the entire area. This advantage will allow you to navigate over multiple routes to any location. It will also preclude your getting lost whenever you miss a turn or are forced off the planned route by obstacles or the tactical situation.

(3) *Resection.* Whenever you have a vantage point to two or more known features portrayed on the map, do not hesitate to use either estimated or plotted resection to pinpoint your position. These opportunities are often plentiful in an urban setting.

This Page intentionally left blank.

CHAPTER 14
UNIT SUSTAINMENT

Land navigation is a skill that is highly perishable. The Soldier must continually make use of the skills he has acquired to remain proficient in them. The institution is responsible for instruction in the basic techniques of land navigation. The institution tests these skills each time a Soldier attends a leadership course. However, it is the unit's responsibility to develop a program to maintain proficiency in these skills between institution courses.

The unit sustainment program provides training that builds on and reinforces the skills the Soldier learned in the institution. It should use the building-block approach to training: basic map reading instruction or review, instruction on land navigation skills, dead reckoning training, dead reckoning practice, terrain association training, terrain association practice, land navigation testing, and building of leader skills. These leader skills should include following a route selected by the commander and planning and following a route selected by the leader.

The unit trainer should be able to set up a sustainment program, a train-the-trainer program, and a land navigation course for his unit's use. It is recommended that units develop a program similar to the one outlined in this chapter. Complete lesson outlines and training plans are available by writing to Commander, 29th Infantry Regiment, ATTN: ATSH-INB-A, Fort Benning, GA 31905-5595.

14-1. SET UP A SUSTAINMENT PROGRAM

The purpose of setting up a sustainment program in the unit is to provide Soldiers with training that reinforces and builds on the training they have received in the institution. All Soldiers should receive this training at least twice a year. The program also provides the unit with a means of identifying the areas in which the Soldiers need additional training.

  a. **Training Guidance.** The unit commander must first determine the levels of proficiency and problems that his unit has in land navigation. This determination can be done through after-action reports from the unit's rotations to NTC/JRTC, ARTEP final reports, feedback from his subordinates, personal observation, and annual training. Once the unit commander decides where his training time should be concentrated, he can issue his training guidance to his subordinate leaders. He also directs his staff to provide training sites, resources, and time for the units to train land navigation. It is recommended that land navigation be trained separately, not just included as a subtask in tactical training.

  b. **Certification.** The unit commander must also provide his subordinate commanders with a means of certifying training. The unit staff must provide subject matter experts to ensure the training meets the standards decided upon by the unit commander. Instructors should be certified to instruct, and courses should be certified before the unit uses it.

  c. **Program Development.** The sustainment program should meet the requirements of all of the unit's Soldiers. It should address all skills from basic map reading to leaders' planning and executing a route. The program should cover the following:
  - Diagnostic examination.

- Map reading instruction/review.
- Land navigation skills training.
- Dead reckoning training/practice.
- Terrain association training/practice.
- Land navigation written/field examination.
- Leaders' training and testing.

The sustainment program should be developed and then maintained in the unit's training files. The program should be developed in training modules so that it can be used as a whole program or used separately by individual modules. It should be designed so the commander can decide which training modules he will use, depending on the proficiency of the unit. The unit commander need only use those modules that fit his training plan.

14-2. SET UP A TRAIN-THE-TRAINER PROGRAM

The purpose of a train-the-trainer program in the unit is to develop trainers capable of providing Soldiers with the confidence and skills necessary to accomplish all assigned land navigation tasks.

    a. **Development of the Program.** The unit commander should appoint a cadre of officers and NCOs to act as primary and alternate instructors for land navigation training. Use the training modules the unit has developed and have these Soldiers go through each module of training until they can demonstrate expertise. Determine which instructors conduct each module of training and have them practice until they are fully prepared to give the training. These instructors act as training cadre for the entire unit. They train their peers to instruct the subordinate units, and they certify each unit's training.

    b. **Conduct of Training.** Conduct training at the lowest level possible. Leaders must be included in all training to keep unit integrity intact.

14-3. SET UP A LAND NAVIGATION COURSE

The unit commander provides specific guidance on what he requires in the development of a land navigation course. It depends upon the unit's mission, training plan, and tasks to be trained. There are basic guidelines to use when setting up a course.

    a. **Determine the Standards.** The unit commander determines the standards for the course. Recommended standards are as follows:

- Distance between points: no less than 300 meters; no more than 1,200 meters.
- Total distance of lanes: no less than 2,700 meters; no more than 11,000 meters.
- Total number of position stakes: no less than seven for each lane; no more than nine for each lane.
- Time allowed: no less than three hours; no more than four hours.

    b. **Decide on the Terrain.** The unit should use terrain that is similar to terrain they will be using in tactical exercises, but, terrain should be different each time training is conducted. The training area for a dismounted course needs to be at least 25 square kilometers; mounted courses require twice as much terrain so that vehicles are not too close to each other.

    c. **Perform a Map and Ground Reconnaissance.** Check the terrain to determine position stake locations, to look for hazards, and to develop training briefings. The following sequence can be used to develop any type of land navigation course. The difference in each course depends on the commander's guidance.

(1) Plot the locations of your position stakes on a 1:50,000-scale map.
(2) Fabricate or order position stakes.
(3) Request support from the local engineer or field artillery unit to survey the locations of your position stakes.
(4) Emplace the position stakes in the surveyed locations.
(5) Certify the course by having your subject matter experts (SMEs) negotiate each lane of the course.
(6) Prepare course requirement sheets and print them.
(7) Complete a risk assessment of the training area.
(8) Begin teaching.

This Page intentionally left blank.

APPENDIX A
SKETCHES

A sketch is a free-hand drawing of a map or picture of an area or route of travel. It shows enough detail and has enough accuracy to satisfy special tactical or administrative requirements.

A-1. PURPOSE
Sketches are useful when maps are not available or the existing maps are not adequate, or to illustrate a reconnaissance or patrol report. Sketches may vary from hasty to complete and detailed, depending upon their purpose and the degree of accuracy required. For example, a sketch of a large minefield will require more accuracy than a hasty sketch of a small unit's defensive position.

A-2. MILITARY SKETCHES
The scale of a sketch is determined by the object in view and the amount of detail required to be shown. The sketch of a defensive position for a platoon or company normally calls for a sketch of larger scale than a sketch for the same purpose for a division. Military sketches also include road and area sketches.

   a. **Field Sketches.** A field sketch (Figure A-1) must show the north arrow, a scale, a legend, and the following features:
   - Power lines.
   - Rivers.
   - Main roads.
   - Towns and villages.
   - Forests.
   - Rail lines.
   - Major terrain features.

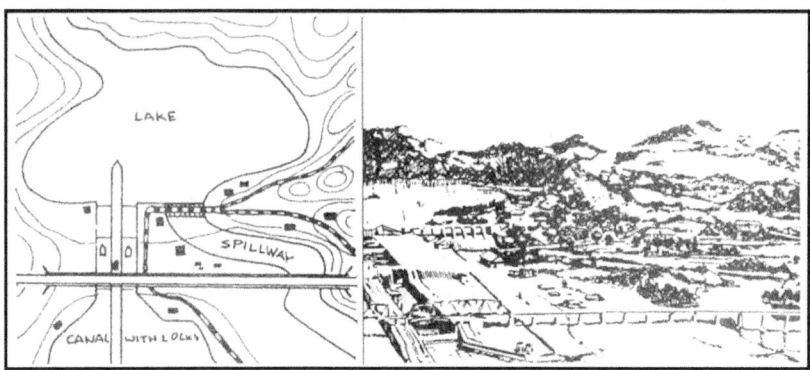

Figure A-1. Sketch map.

b. **Road Sketches**. Road sketches show the natural and military features on and in the immediate vicinity of the road. In general, the width of terrain sketches will not exceed 365 meters on each side of the road. Road sketches may be used to illustrate a road when the existing map does not show sufficient detail.

c. **Area Sketches**. Area sketches include sketches of positions, OPs, or particular places.

(1) *Position Sketch*. A position sketch is one of a military position, campsite, or other area of ground. To effectively complete a position sketch, the sketcher must have access to all parts of the area being sketched.

(2) *Observation Post Sketch*. An OP sketch shows the military features of ground along a friendly OP line as far toward the enemy position as possible.

(3) *Place Sketch*. A place sketch is one of an area made by a sketcher from a single point of observation. Such a sketch may cover ground in front of an OP line, or it may serve to extend a position or road sketch toward the enemy.

FM 3-25.26

APPENDIX B
MAP FOLDING TECHNIQUES

One of the first considerations in the care of maps is proper folding.

B-1. FOLDING METHODS
Figure B-1 shows two ways of folding maps to make them small enough to be carried easily and still be available for use without having to unfold them entirely.

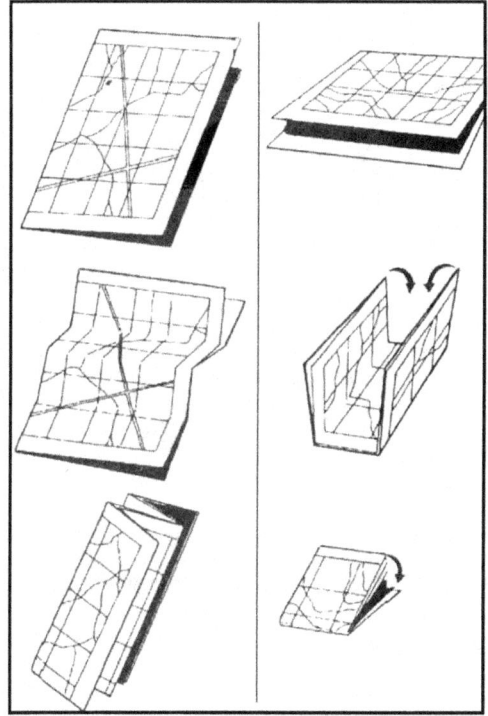

Figure B-1. Two methods of folding a map.

FOUO
18 January 2005

B-1

B-2. PROTECTION OF MAP

After a map has been folded, it should be placed in a folder for protection. Apply adhesive to the back of the segments corresponding to A, F, L, and Q (Figure B-2).

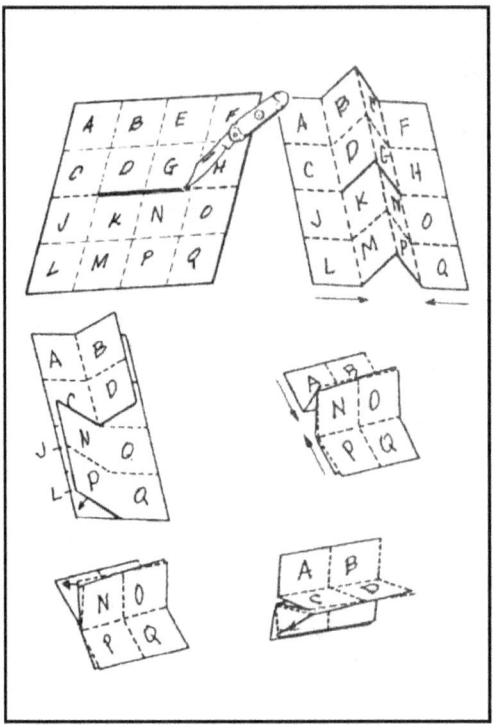

Figure B-2. How to slit and fold a map for special use.

NOTE: Before attempting to cut and fold a map as illustrated in Figure B-2, make a practice cut and fold with a piece of paper.

APPENDIX C
UNITS OF MEASURE AND CONVERSION FACTORS

This appendix provides conversion tables for units of measure and conversion factors that are used in military operations (Tables C-1 through C-5, pages C-1 through C-3).

12 inches	=	1 foot
36 inches	=	1 yard
3 feet	=	1 yard
1,760 yards	=	1 mile statute
2,026.8 yards	=	1 mile nautical
5,280 feet	=	1 mile statute
6,080.4 feet	=	1 mile nautical
63,360 inches	=	1 mile statute
72,963 inches	=	1 mile nautical

Table C-1. English system of linear measure.

1 millimeter	=	0.1 centimeter	=	0.0393 inches
10 millimeters	=	1.0 centimeter	=	0.3937 inches
10 centimeters	=	1.0 decimeter	=	3.937 inches
10 decimeters	=	1.0 meter	=	39.37 inches
10 meters	=	1.0 decameter	=	32.81 feet
10 decameters	=	1.0 hectometer	=	328.1 feet
10 hectometers	=	1.0 kilometer	=	0.62 mile
10 kilometers	=	1.0 myriameter	=	6.21 miles

Table C-2. Metric system of linear measure.

1 mil	=	1/6400 circle	=	0.05625°	=	0.0625 grad
1 grad	=	1/400 circle	=	16.0 mils	=	0°54' = 0.9°
1 degree	=	1/360 circle	=	about 17.8 mils	=	about 1.1 grad

Table C-3. Equivalent units of angular measure.

ONE	INCHES	FEET	YARDS	STATUTE MILES	NAUTICLE MILES	mm
Inch	1	0.0833	0.0277	-	-	25.40
Foot	12	1	0.333	-	-	304.8
Yard	36	3	1	0.00056	-	914.4
Statute Mile	63,360	5,280	1,760	1	0.8684	-
Nautical Mile	72,963	6,080	2,026	1.1516	1	-
Millimeter	0.0394	0.0033	0.0011	-	-	1
Centimeter	0.3937	0.0328	0.0109	-	-	10
Decimeter	3.937	0.328	0.1093	-	-	100
Meter	39.37	3.2808	1.0936	0.0006	0.0005	1,000
Decameter	393.7	32.81	10.94	0.0062	0.0054	10,000
Hectometer	3,937	328.1	109.4	0.0621	0.0539	100,000
Kilometer	39,370	3,281	1,094	0.6214	0.5396	1,000,000
Myriameter	393,700	32,808	10,936	6.2137	5.3959	10,000,000

ONE	cm	dm	M	dkm	hm	km	mym
Inch	2.540	0.2540	0.0254	0.0025	0.0003	-	-
Foot	30.48	3.048	0.3048	0.0305	0.0030	0.0003	-
Yard	91.44	9.144	0.9144	0.0914	0.0091	0.0009	-
Statute Mile	160,930	16,093	1,609	160.9	16.09	1.6093	0.1609
Nautical Mile	185,325	18,532	1,853	185.3	18.53	1.8532	0.1853
Millimeter	0.1	0.01	0.001	0.0001	-	-	-
Centimeter	1	0.1	0.01	0.001	0.0001	-	-
Decimeter	10	1	0.1	0.01	0.001	0.0001	-
Meter	100	1	1	0.1	0.01	0.001	0.0001
Decameter	1,000	10	10	1	0.1	0.01	0.001
Hectometer	10,000	100	100	10	1	0.1	0.01
Kilometer	100,000	1,000	1,000	100	10	1	0.1
Myriameter	1,000,000	10,000	10,000	1000	100	10	1

Table C-4. Conversion factors.

Example I:
    Problem: Reduce 76 centimeters to (?) inches.
    76 cm x 0.3937 = 29 inches
    Answer: There are 29 inches in 76 centimeters.

Example II:
    Problem: How many feet are there in 2.74 meters?

$$\frac{2.74}{.3048} = 9 \text{ feet}$$

    Answer: There are approximately 9 feet in 2.74 meters.

SCALE	1 INCH EQUALS	1 CENTIMETER EQUALS
1:5,000	416.67 feet 127.00 meters	164.00 feet 50.00 meters
1:10,000	833.33 feet 254.00 meters	328.10 feet 100.00 meters
1:12,500	1,041.66 feet 317.00 meters	410.10 feet 125.00 meters
1:20,000	1,666.70 feet 508.00 meters	656.20 feet 200.00 meters
1:25,000	2,083.30 feet 635.00 meters	820.20 feet 250.00 meters
1:50,000	4,166.70 feet 1,270.00 meters	1,640.40 feet 500.00 meters
1:63,360	5,280.00 feet 1,609.30 meters	2,078.70 feet 633.60 meters
1:100,000	8,333.30 feet 2,540.00 meters	3,280.80 feet 1,000.00 meters
1:250,000	20,833.00 feet 6,350.00 meters	8,202.00 feet 2,500.00 meters
1:500,000	41,667.00 feet 12,700.00 meters	16,404.00 feet 5,000.00 meters

Table C-5. Ground distance at map scale.

This Page intentionally left blank.

APPENDIX D
JOINT OPERATIONS GRAPHICS

Joint operations graphics (Chapter 2, paragraph 2-6b[4]) are based on the format of the standard 1:250,000-scale military topographic maps. They contain additional information needed in present-day joint air-ground operations.

D-1. TYPES
Each JOG is prepared in two types; one is designed for air operations and the other for ground operations. Each version is identified in the lower margin as JOINT OPERATIONS GRAPHIC (AIR) or JOINT OPERATIONS GRAPHIC (GROUND).

D-2. BASIC CONTENTS
The basic topographic information is the same on both JOG versions.
  a. Power transmission lines are symbolized as a series of purple pylons connected by a solid purple line.
  b. Airports, landing facilities, and related air information are shown in purple. The purple symbols that may be unfamiliar to the user are shown in the legend in the margin.
  c. The top of each obstruction to air navigation is identified by its elevation above sea level and its elevation above ground level.
  d. Along the north and east edges of the graphic, detail is extended beyond the standard sheet lines to create an overlap with the graphics to the north and to the east.
  e. Layer tinting (paragraph 10-2a) and relief shading (paragraph 10-2c) are added as an aid to interpreting the relief.
  f. The incidence of the graphic in the world geographic reference system (paragraph 4-8b) is shown by a diagram in the margin.

D-3. JOINT OPERATIONS GRAPHIC (AIR)
The JOG (AIR) series, prepared for air use, contains detailed information on air facilities such as radio ranges, runway lengths, and landing surfaces. The highest terrain elevation in each 15-minute quadrangle is identified by the large open-faced figures shown in the legend. Elevations and contours on JOG (AIR) sheets are given in feet.

D-4. JOINT OPERATIONS GRAPHIC (GROUND)
The JOG (GROUND) series is prepared for use by ground units, and only stable or permanent air facilities are identified. Elevations and contours are located in the same positions as on the air version, but are given in meters.

This Page intentionally left blank.

APPENDIX E
ORIENTEERING

What is orienteering? Orienteering is a competitive form of land navigation suitable for all ages and degrees of fitness and skill. It provides the suspense and excitement of a treasure hunt. The object of orienteering is to locate control points by using a map and compass to navigate through the woods. The courses may be as long as 10 kilometers.

E-1. HISTORY

Orienteering began in Scandinavia in the nineteenth century. It was primarily a military event and was part of military training. Not until 1919 was the modern version of orienteering born in Sweden as a competitive sport. Ernst Killander, its creator, can be rightfully called the father of orienteering. In the early thirties, the sport received a technical boost with the invention of a new compass, more precise and faster to use. The Kjellstrom brothers, Bjorn and Alvan, and their friend, Brunnar Tillander, were responsible for this new compass. They were among the best Swedish orienteers of the thirties, with several individual championships among them. Orienteering was brought into the U.S. in 1946 by Bjorn Kjellstrom.

E-2. DESCRIPTION

Each orienteer is given a 1:50,000 topographic map with the various control points circled. Each point has a flag marker and a distinctive punch that is used to mark the scorecard. Competitive orienteering involves running from checkpoint to checkpoint. It is more demanding than road running, not only because of the terrain, but because the orienteer must constantly concentrate, make decisions, and keep track of the distance covered. Orienteering challenges both the mind and the body; however, the competitor's ability to think under pressure and make wise decisions is more important than speed or endurance.

E-3. THE COURSE

The orienteering area should be on terrain that is heavily wooded, preferably uninhabited, and difficult enough to suit different levels of competition. The area must be accessible to competitors and its use must be coordinated with appropriate terrain and range control offices.

  a. The ideal map for an orienteering course is a multicolored, accurate, large-scale topographic map. A topographic map is a graphic representation of selected man-made and natural features of a part of the earth's surface plotted to a definite scale. The distinguishing characteristic of a topographic map is the portrayal of the shape and elevation of the terrain by contour lines.

  b. For orienteering within the United States, large-scale topographic (topo) maps are available from the National Geospatial-Intelligence Agency (NGA) Hydrographic Topographic Center. The scale suitable for orienteering is 1:50,000.

E-4. SETTING UP THE COURSE

The challenge for the course setter is to keep the course interesting, but never beyond the individual's or group's ability. General guidance is to select locations that are easily identifiable on the map and terrain, and accessible from several routes.

   a. Those who set up the initial event should study a map for likely locations of control points and verification of the locations. Better yet, they should coordinate with an experienced competitor in selecting the course.

   b. Orienteering includes several forms of events. Some of the most common are route, line, cross-country, and score orienteering.

      (1) *Route Orienteering*. This form can be used during the training phase and in advanced orienteering.

         (a) In this type of event, a master or advanced competitor leads the group as they walk a route. Beginners trace the actual route walked on the ground onto their maps. They circle the location of the different control points found along the walked route. When they finish, the maps are analyzed and compared. During training, time is not a factor.

         (b) Another variation is when a course is laid out on the ground with markers for the competitor to follow. There is no master map, as the course is traced for the competitor by flags or markers. The winner of the event is the competitor who has successfully traced the route and accurately plotted the most control points on his map.

      (2) *Line Orienteering*. At least five control points are used during this form of orienteering training. The competitor traces on his map a preselected route from a master map. The object is to walk the route shown on the map, circling the control points on the map as they are located on the ground (Figure E-1).

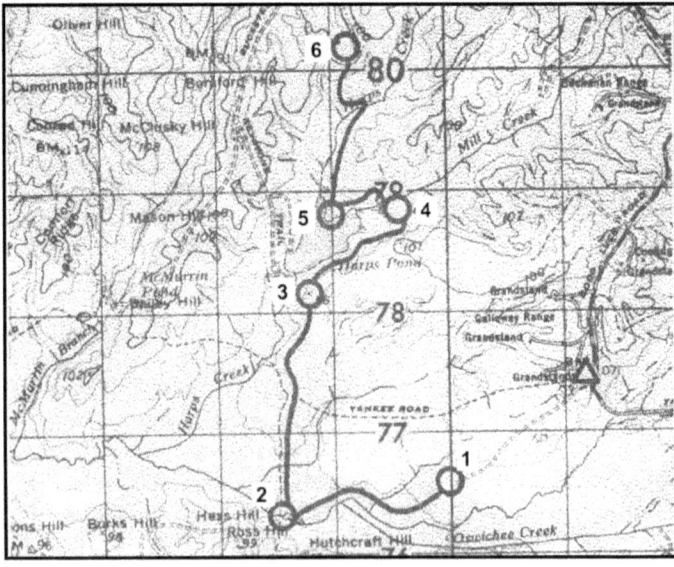

Figure E-1. Line orienteering.

(3) ***Cross-Country Orienteering***. This is the most common type of orienteering competitions. It is sometimes called free or point orienteering and is considered to be the most competitive and intriguing of all events (Figure E-2). In this event, all competitors must visit the same controls in the same order. With the normal one-minute starting interval, it becomes a contest of route choice and physical skill. The winner is the contestant with the fastest time around the course.

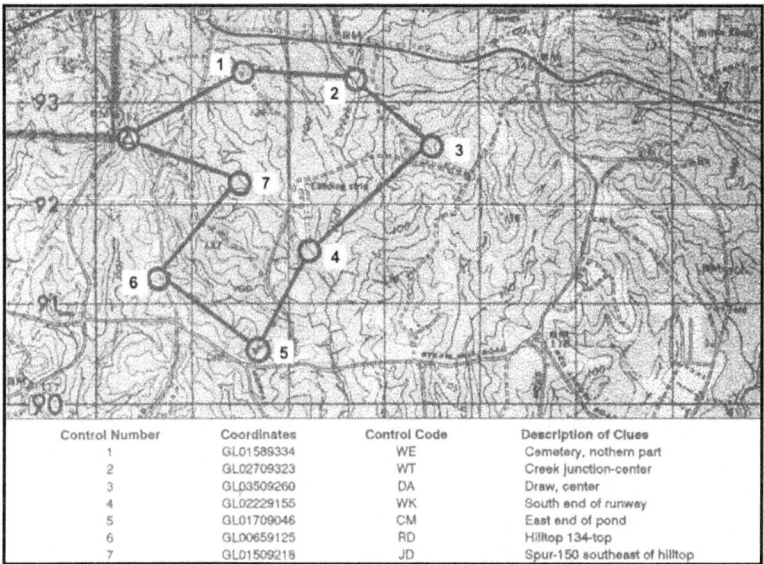

Figure E-2. A cross-country orienteering map.

(a) After selecting the control points for the course, determine the start and finish locations. The last control point should be near the finish. In describing each control's location, an eight-digit grid coordinate and a combination of two letters identifying the point (control code) should be included in each descriptive clue list that is normally given to each competitor at least two minutes before his start time.

(b) There are usually 6 to 12 control markers on the course in varying degrees of difficulty and distances apart so that there are no easy, direct routes. Instead, each competitor is faced with many choices of direct but difficult routes, or of indirect but easier routes. Each control's location is circled, and the order in which each is to be visited is clearly marked on the master map. The course may be a closed transverse with start and finish collocated, or the start and finish may be at different locations. The length of the course and difficulty of control placement varies with the competitors' degree of expertise. Regardless of the class of event, all competitors must indicate on their event cards proof of visiting the control markers. Inked stamps, coded letters, or punches are usually used to do this procedure.

FM 3-25.26

NOTE: The same orienteering range may serve in both cross-country and score events. However, a separate set of competitor maps, master maps, and event cards are necessary.

(4) *Score Orienteering.* In this event, the area chosen for the competition is blanketed with many control points (Figure E-3). The controls near the start/finish point (usually identical in this event) have a low point value, while those more distant or more difficult to locate have a high point value. This event requires the competitor to locate as many control markers as he can within the specified time (usually 90 minutes). Points are awarded for each control visited and deducted for exceeding the specified time. The competitor with the highest point score is the winner.

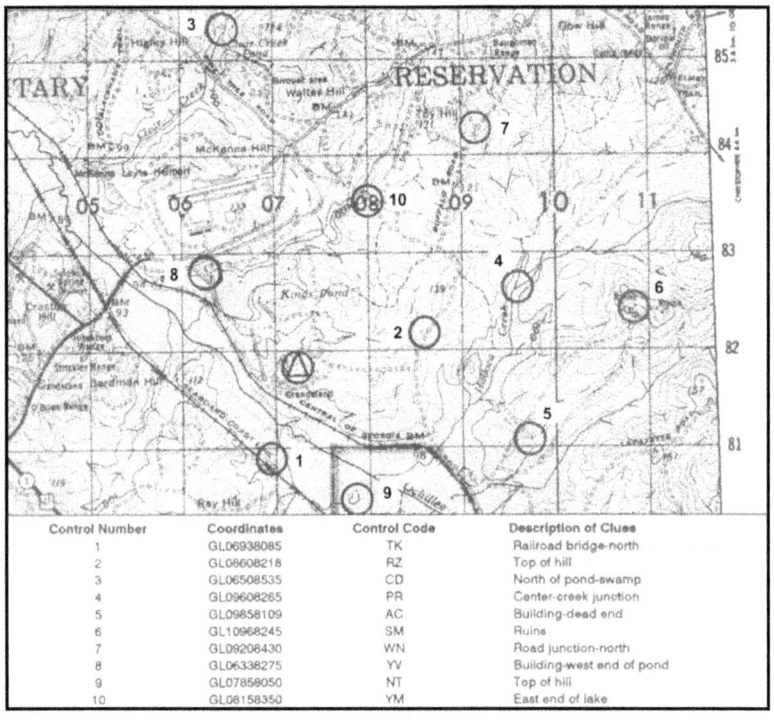

Figure E-3. A score orienteering map.

(a) Conducting a score event is basically the same as the cross-country event at the start. The competitor is given a map and an event card. The event card lists all the controls with their different point values. When released to the master map, the competitor finds the circles and numbers indicating the location of all the controls listed on his event card. He copies all the red circles on his map. Then he chooses any route he wishes to take in amassing the highest possible point score in the time available. The course is designed to ensure that there

E-4 FOUO
18 January 2005

are more control points than can possibly be visited in the allotted time. Again, each control marker visited must be indicated on the event card.

(b) It is important for the competitor to take time initially to plot the most productive route. A good competitor may spend up to 6 minutes in the master map area while plotting the ideal route.

(c) There is no reward for returning early with time still available to find more points, so the good competitor must be able to coordinate time and distance with his ability in land navigation in running the course.

E-5. OFFICIALS

The same officials can be used at the start and finish. Although more officials or assistants can be used, the following paragraphs list the minimum that can be used for a competition.

    a. **At the Start.** These officials or assistants are needed at the start.

    (1) *Course Organizer*—Briefs the orienteers in the assembly area, issues event cards and maps, and calls orienteers forward to start individually.

    (2) *Recorder*—Records orienteer's name and start time on recorder's sheet, checks orienteer's name and start number on his event card, and issues any last-minute instructions.

    (3) *Timer*—Controls the master clock and releases the orienteers across the start line at their start time (usually at one-minute intervals) to the master map area.

    b. **At the Finish.** These officials or assistants are needed at the finish.

    (1) *Timer*—Records finish time of each orienteer on the orienteer's event card and passes the card to the recorder.

    (2) *Recorder*—Records finish time of each orienteer on the orienteer's event card and passes the card to the course organizer.

    (3) *Course Organizer*—Verifies correctness of names, finish times, and final score; posts orienteers' positions on results board; and accounts for all orienteers at the end of event.

E-6. START/FINISH AREA

The layout of the start/finish areas for orienteering events is basically the same for all forms.

    a. **Assembly Area.** This is where orienteers register and receive instructions, maps, event cards, and start numbers. They may also change into their orienteering clothes if facilities are available, study their maps, and fill out their event cards here. Sanitation facilities should be available in this area.

    b. **Start.** At the start, the orienteer reports to the recorder's and timer's table to be logged in by the recorder and released by the timer.

    c. **Master Map Area.** There are three to five master maps 20 to 50 meters from the start. When the orienteer arrives at this area, he must mark his map with all the course's control points. Having done this, he must decide on the route that he is to follow. A good orienteer takes the time to orient his map and carefully plot his route before rushing off. It is a good idea to locate the master map area out of sight of the start point to preclude orienteers tracking one another.

    d. **Equipment.** The following is a list of equipment needed by the host of an orienteering event:

- Master maps, three to five, mounted.

FM 3-25.26

- Competitor maps, one each.
- Event cards, one each.
- Recorder's sheets, two.
- Descriptive clue cards, one each.
- Time clocks, two.
- Rope, 100 to 150 feet, with pegs for finish tunnel.
- Card tables, one or two.
- Folding chairs, two or three.
- Results board.
- Control markers, one per point.
- Extra compasses.
- Whistle, for starting.
- First aid kit.
- Colored tape or ribbon for marking route to master map and from last control point to finish.

e. **Control Markers.** These are orange-and-white markers designating each control point (Figure E-4). Ideally, they should have three vertical square faces, forming a triangle with the top and bottom edges. Each face should be 12 inches on a side and divided diagonally into red and white halves or cylinders (of similar size) with a large, white, diagonal stripe dividing the red cylinder. For economy or expediency, 1-gallon milk cartons, 5-gallon ice cream tubs, 1-gallon plastic bleach bottles, or foot-square plaques, painted in the diagonal or divided red and white colors of orienteering, may be used.

Figure E-4. Control markers.

(1) Each marker should have a marking or identification device for the orienteer to use to indicate his visit to the control. This marker may be the European-style punch pliers, a self-inking marker, different colored crayons at each point, different letter combinations, different number combinations, or different stamps or coupons. The marking device must be unique, simple, and readily transcribable to the orienteers' event cards.

(2) The control marker should normally be visible from at least 10 meters. It should not be hidden.

f. **Recorder's Sheets.** A suggested format for the recorder's sheet is depicted in Figure E-5.

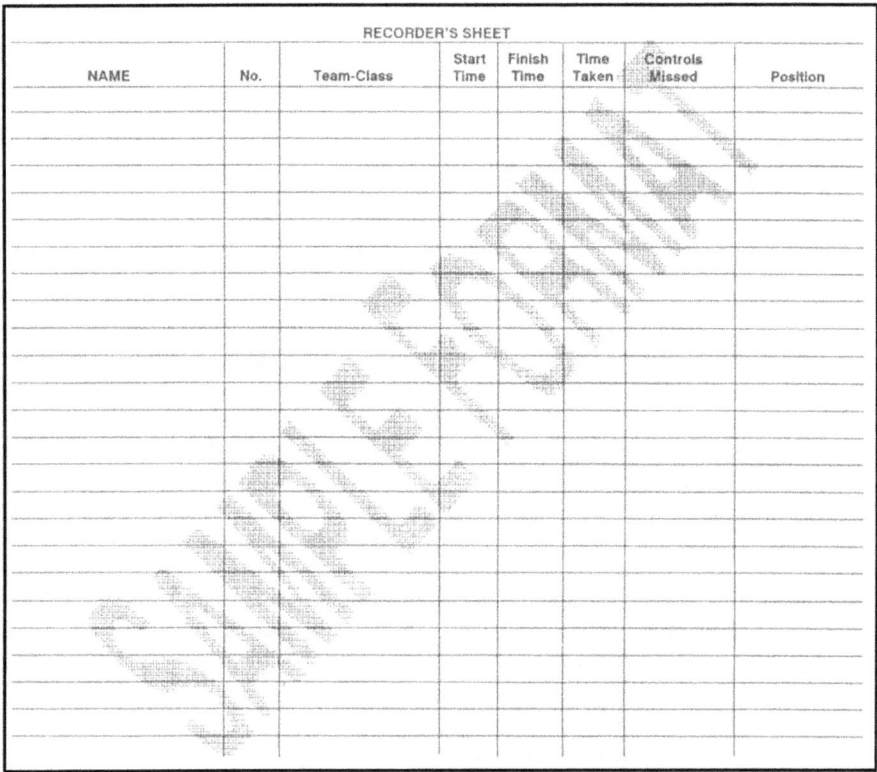

Figure E-5. Sample recorder's sheet.

FM 3-25.26

g. **Event Card.** The event card can be made before the event and should be as small as possible, as it is carried by the competitor. It must contain the following items: name, start number, start time, finish time, total time, place, and enough blocks for marking the control points. As indicated earlier, it may also contain a listing of descriptive clues (Figure E-6).

```
+---------------------------------------------------------------+
|              CROSS-COUNTRY ORIENTEERING TEAM                  |
|                                                               |
| NAME _____ COMPANY _____ COURSE _____ TEAM _____    |
| NAME _____ COMPANY _____ START TIME ___ FINISH TIME __  |
| CHECKPOINTS           DESCRIPTION CLUES                       |
|  +---+---+                                                    |
|  | 1 | 2 |                                                    |
|  +---+---+                                                    |
|  | 3 | 4 |                                                    |
|  +---+---+                                                    |
|  | 5 | 6 |   NOTE: All control signes are located at eye      |
|  +---+---+         level on trees.                            |
|                              1. All work is individual team   |
| Total Value of Points ____      effort.                       |
|                              2. You must not join with or     |
| Penalty Points _____      coordinate with any other     |
|                                 team.                         |
| Final Score _____   3. You must personally visit     |
|                                 each point you indicate on    |
|                                 your scorecard.               |
+---------------------------------------------------------------+
```

Figure E-6. Sample cross-country orienteering event card.

h. **Results Board.** This board displays the orienteer's position in the event at the finish (Figure E-7). There are a variety of ways of displaying the results, from blackboard to ladder-like to a clothesline-type device where each orienteer's name, point score, and times are listed.

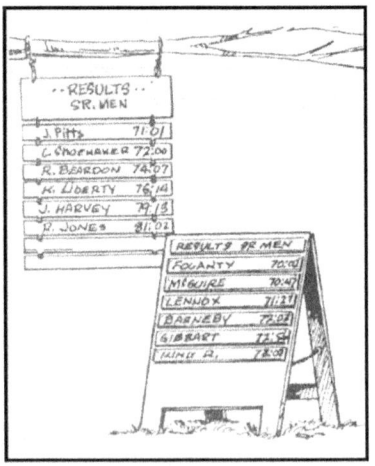

Figure E-7. Sample results board.

E-8 FOUO
18 January 2005

i. **Clue Description Card.** These cards are prepared with the master maps after the course is set. They contain the descriptive clues for each control point, control code, grid coordinate references, returning time for competitors, removal times for each location, and panic azimuth (Figure E-8). The terminology on these must be identical to that listed in the definition section. These cards and the master maps must be kept confidential until the orienteers start the event.

Figure E-8. Sample clue description card.

j. **Scoring.** The cross-country or free event is scored by the orienteer's time alone. All control points must be visited; failure to visit one results in disqualification. In this event, the fastest time wins.

(1) A variation that can be introduced for novices is to have a not-later-than return time at the finish and add minutes to the orienteer's final time for minutes late and control points not located.

(2) The score event requires the amassing of as many points as possible within the time limit. Points are deducted for extra time spent on the course, usually one point for each 10 seconds extra.

k. **Prizes.** A monetary prize is not awarded. A suggested prize for beginners is an orienteering compass or some other practical outdoor-sports item.

E-7. SAFETY ON THE COURSE

A first aid kit must be available at the start and finish. One of the officials should be trained in first aid or have a medic at the event. Other safety measures include:

a. **Control Points.** Locate the controls where the safety of the competitor is not jeopardized by hazardous terrain or other circumstances.

b. **Safety Lane.** Have a location, usually linear, on the course where the competitor may go if injured, fatigued, or lost. A good course will usually have its boundary as a safety lane. Then a competitor can set a panic azimuth on the compass and follow it until he reaches the boundary.

c. **Finish Time.** All orienteering events must have a final return time. At this time, all competitors must report to the finish line even if they have not completed the course.

d. **Search-and-Rescue Procedures.** If all competitors have not returned by the end of the competition, the officials should drive along the boundaries of the course to pick up the missing orienteers.

E-8. CONTROL POINT GUIDELINES

When the control point is marked on the map as well as on the ground, the description of that point is prefaced by the definite article *the*; for example, *the pond*. When the control point is marked on the ground but is not shown on the map, then the description of the point is prefaced by the indefinite article *a*; for example, *a trail junction*. In this case, care must be taken to ensure that no similar control exists within at least 25 meters. If it does, then either the control must not be used or it must be specified by a directional note in parentheses; for example, a depression (northern). Other guidelines include:

a. Points of the compass are denoted by capital letters; for example, S, E, SE.

b. Control points within 100 meters of each other or different courses are not to be on the same features or on features of the same description or similar character.

c. For large (up to 75 meters across) features or features that are not possible to see across, the position of the control marker on the control point should be given in the instructions. For example, the east side of the pond; the north side of the building.

d. If a very large (100 to 200 meters) feature is used, the control marker should be visible from most directions from at least 25 meters.

e. If a control point is near but not on a conspicuous feature, this fact and the location of the marker should be clearly given; for example, 10 meters E of the junction. Avoid this kind of control point.

f. Use trees in control descriptions only if they are prominent and a totally different species from those surrounding. Never use bushes and fauna as control points.

g. Number control points in red on the master map.

h. For cross-country events, join all control points by a red line indicating the course's shape.

E-9. MAP SYMBOLS

The map symbols in Figure E-9 are typical topographic and cultural symbols that can be selected for orienteering control points. The map cutouts have been selected from DMA maps.

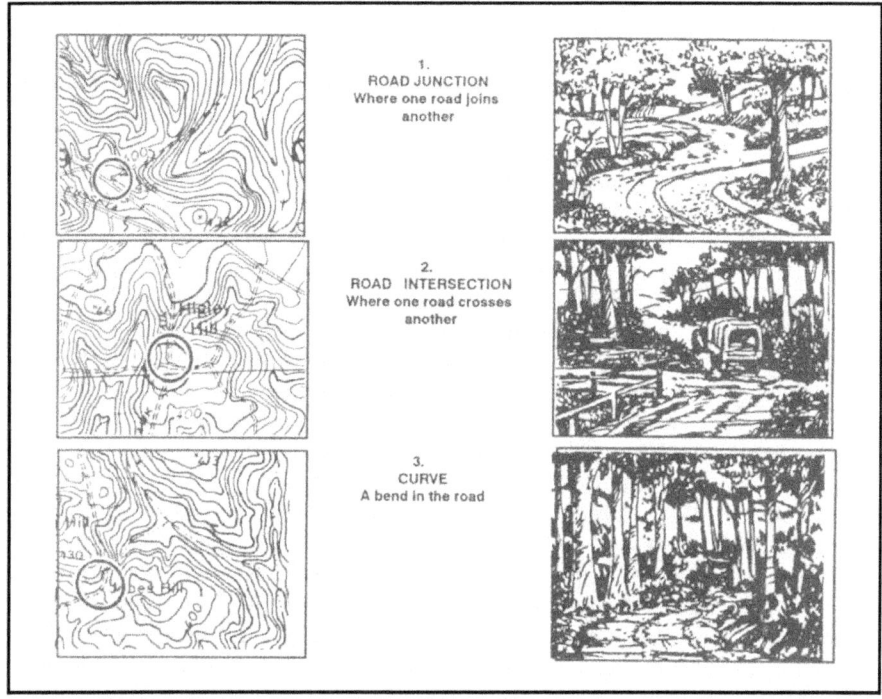

Figure E-9. Map symbols.

Figure E-9. Map symbols (continued).

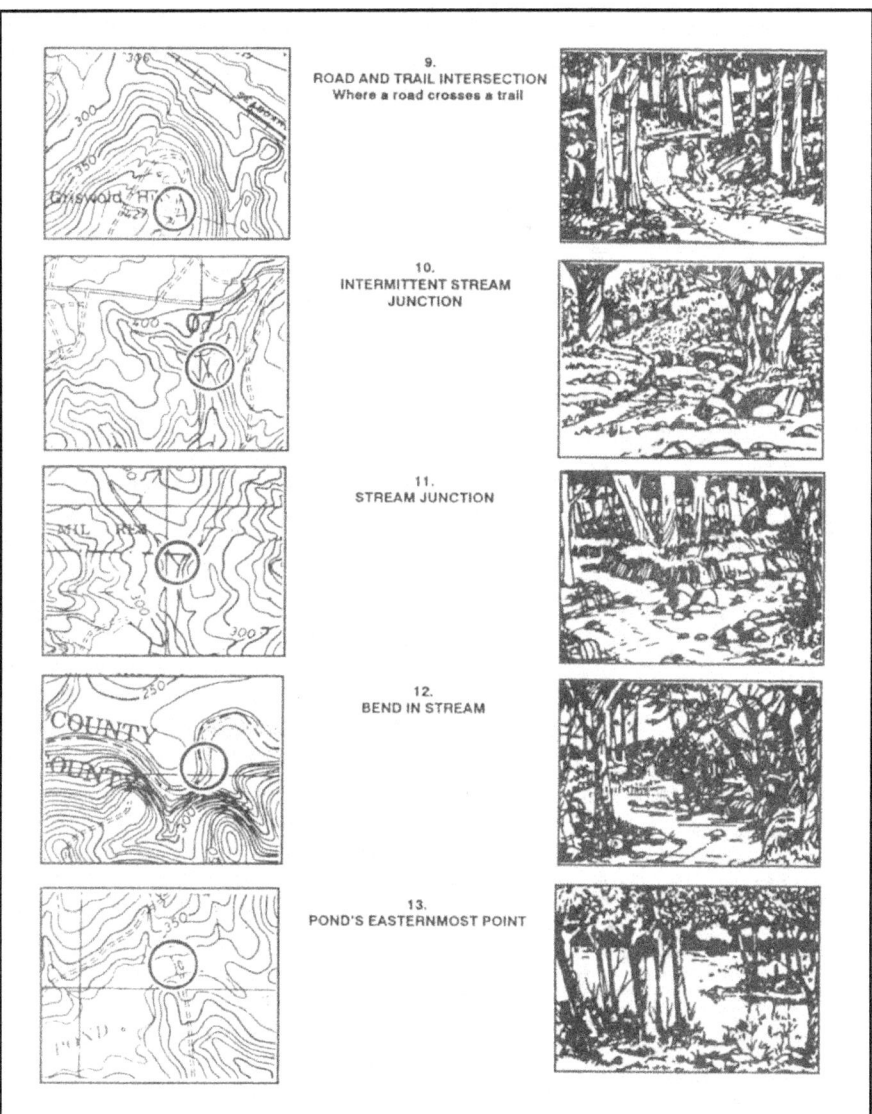

Figure E-9. Map symbols (continued).

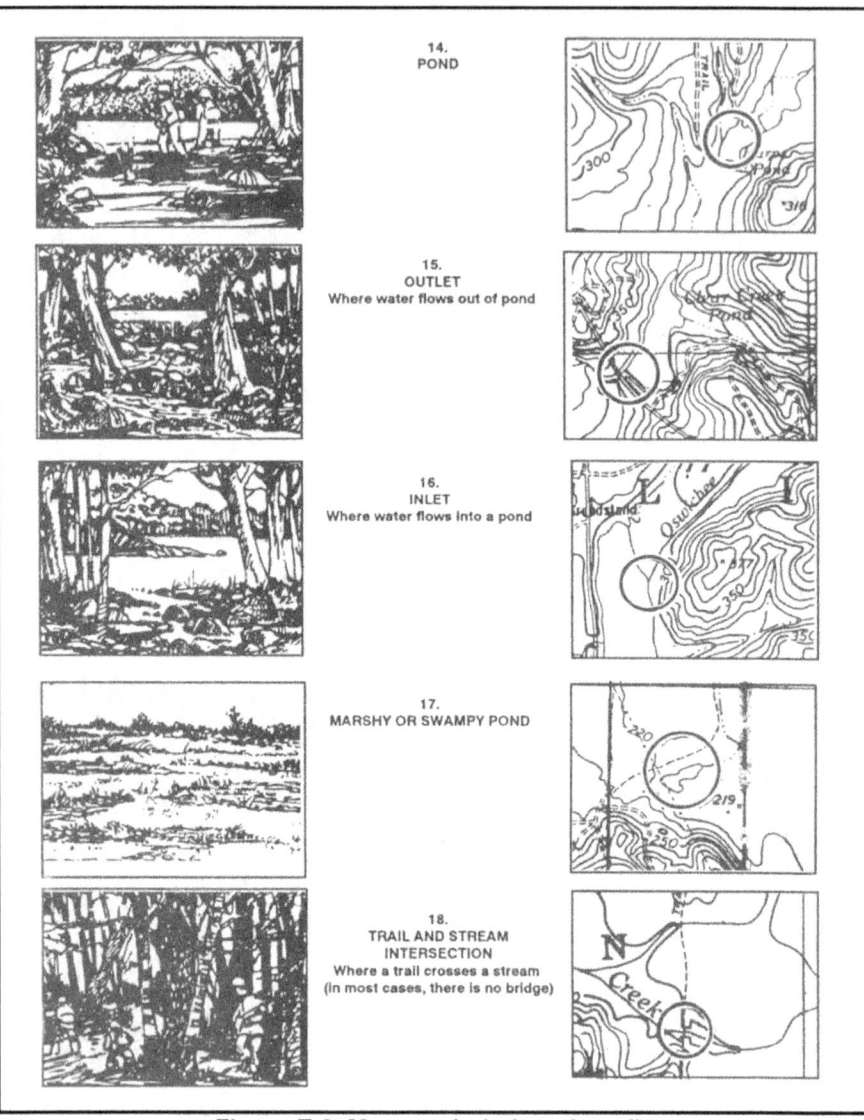

Figure E-9. Map symbols (continued).

Figure E-9. Map symbols (continued).

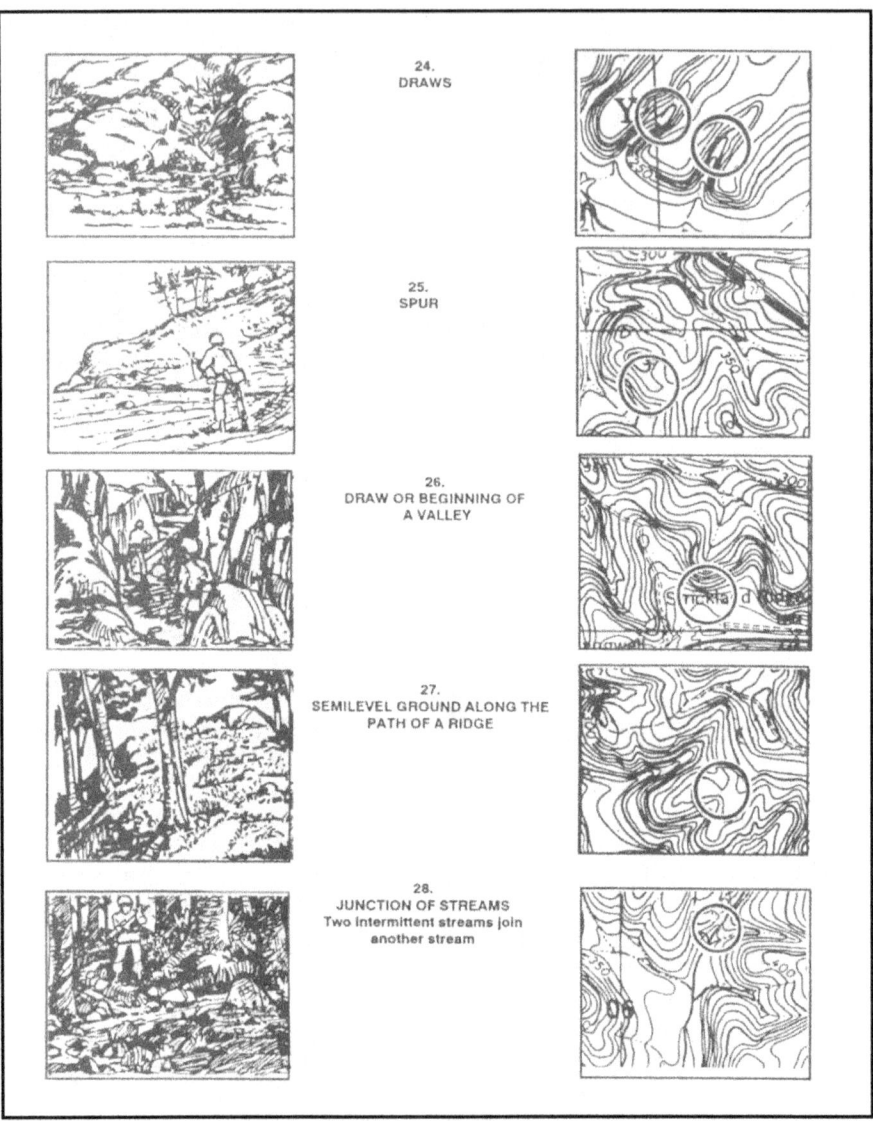

Figure E-9. Map symbols (continued).

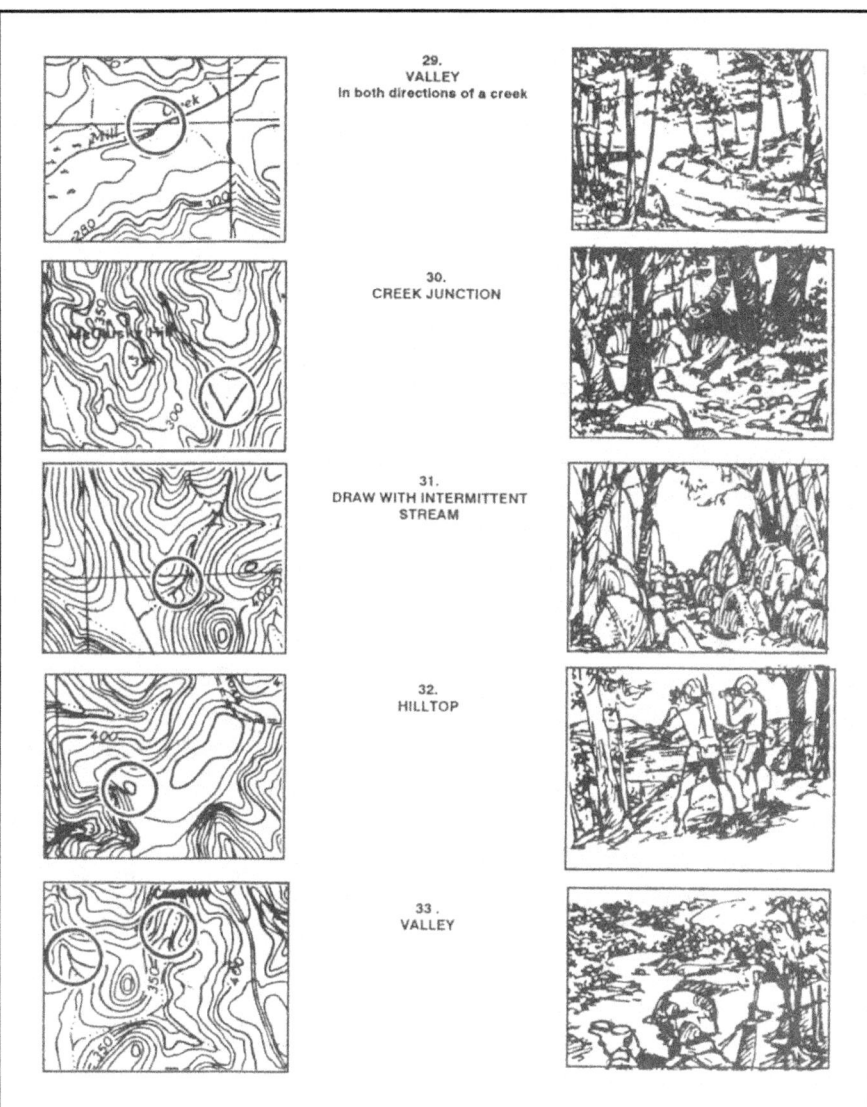

Figure E-9. Map symbols (continued).

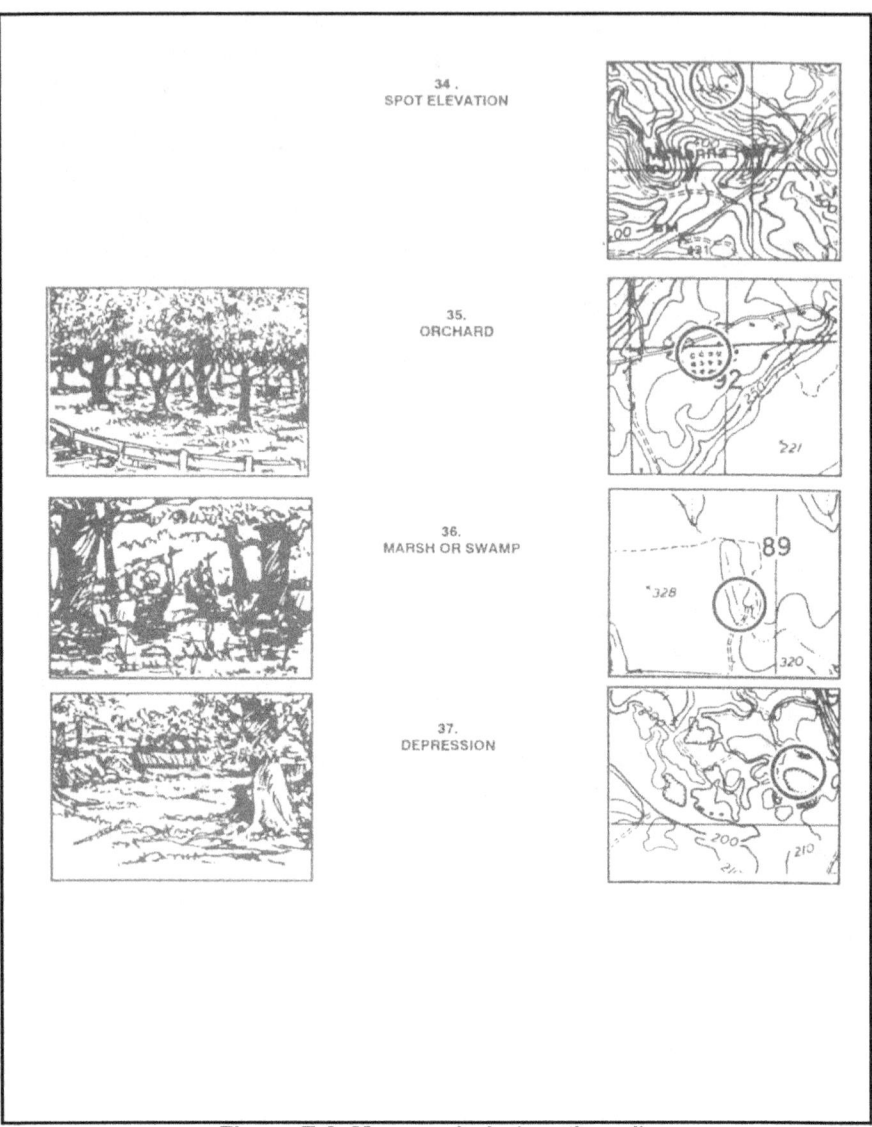

Figure E-9. Map symbols (continued).

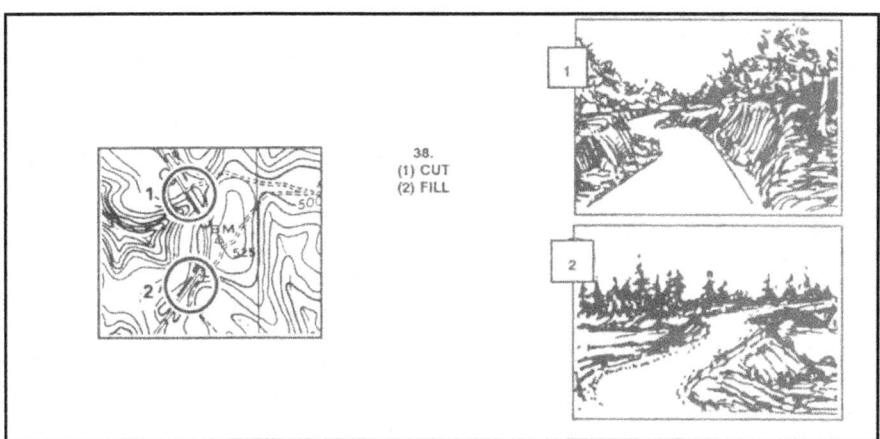

Figure E-9. Map symbols (continued).

E-10. ORIENTEERING TECHNIQUES

The orienteer should try not to use the compass to orient the map. The terrain association technique is recommended instead. The orienteer should learn the following techniques:

a. **Pacing.** One of the basic skills that the orienteer should develop early is how to keep track of distance traveled while walking and running. This is done on a 100-meter pace course.

b. **Thumbing.** This technique is very simple, but the map has to be folded small to use it. The orienteer finds his location on the map and places his thumb directly next to it. He moves from point to point on the ground without moving his thumb from his initial location. To find the new location, the only thing that he has to do is look at the map and use his thumb as a point of reference for his last location. This technique prevents the orienteer from looking all over the map for his location.

c. **Handrails.** This technique enables the orienteer to move rapidly on the ground by using existing linear features (such as trails, fences, roads, and streams) that are plotted along his route. They can also be used as limits or boundaries between control points (Figure E-10, page E-20).

d. **Attack Points.** These are permanent known landmarks that are easily identified on the ground. They can be used as points of reference to find control points located in the woods. Some examples of attack points are stream junctions, bridges, and road intersections.

FM 3-25.26

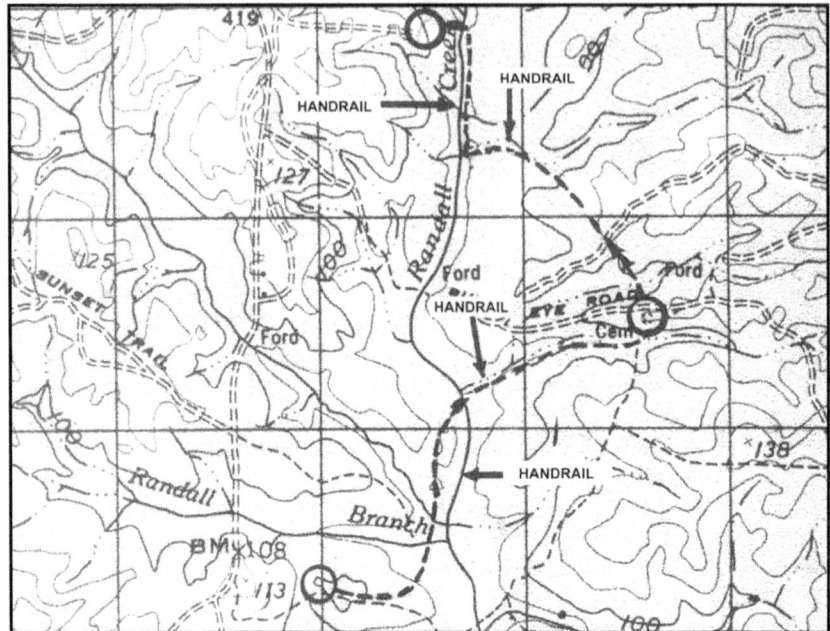

Figure E-10. Handrails.

E-11. CIVILIAN ORIENTEERING

Civilian orienteering is conducted under the guidelines of the United States Orienteering Federation with at least 70 clubs currently affiliated. Although civilian orienteering is a form of land navigation, the terms, symbols, and techniques are different from the military.

  a. An expert military map reader/land navigator is by no means ready to compete in a civilian orienteering event. However, military experience in navigating on the ground and reading maps will help individuals to become good orienteers. Several orienteering practices and complete familiarization with the map symbols and terms before participating in a real orienteering event is recommended.

  (1) *Map*. The standard orienteering map is a very detailed, 1:15,000-scale, colored topographical map. All orienteering maps contain only north-south lines that are magnetically drawn; this eliminates any declination conversions. Because of the absence of horizontal lines, grid coordinates cannot be plotted and therefore are not needed.

  (2) *Symbols (Legend)*. Despite standard orienteering symbols, the legend in orienteering maps has a tendency to change from map to map. A simple way to overcome this problem is to become familiar with the legend every time a different map is used.

  (3) *Scale*. The scale of orienteering maps is 1:15,000. This requires an immediate adjustment for the military land navigator, especially while moving from point to point. It takes a while for a person that commonly uses a 1:50,000 scale to get used to the orienteering map.

  (4) *Contours*. The normal contour interval in an orienteering map is 5 meters. This interval, combined with the scale, makes the orienteering maps so meticulously detailed that

a 1-meter boulder, a 3-meter shallow ditch, or a 1-meter depression will show on the map. This may initially shock a new orienteer.

(5) ***Terms and Description of Clues.*** The names of landforms are different from those commonly known to the military. For example, a valley or a draw is known as a reentrant; an intermittent stream is known as a dry ditch. These terms, with a description of clues indicating the position and location of the control points, are used instead of grid coordinates.

b. The characteristics of the map, the absence of grid coordinates, the description of clues, and the methods used in finding the control points are what make civilian orienteering different from military land navigation.

This Page intentionally left blank.

FM 3-25.26

APPENDIX F
M2 COMPASS

The M2 compass is a rustproof and dustproof magnetic instrument that provides slope, angle of site, and azimuth readings. One of the most important features of the M2 compass is that it is graduated in mils and does not require a conversion from degrees to mils as does the M1 compass. It can be calibrated to provide a grid azimuth or it can be used without calibration to determine a magnetic azimuth.

The M2 compass (Figure F-1; Figure F-2, page F-3; and Figure F-3, page F-4) is a multiple-purpose instrument used primarily to obtain azimuths and angles of site. It also measures grid azimuths after the instrument has been declinated for the locality. (For more detailed information, see TM 9-1290-333-15.)

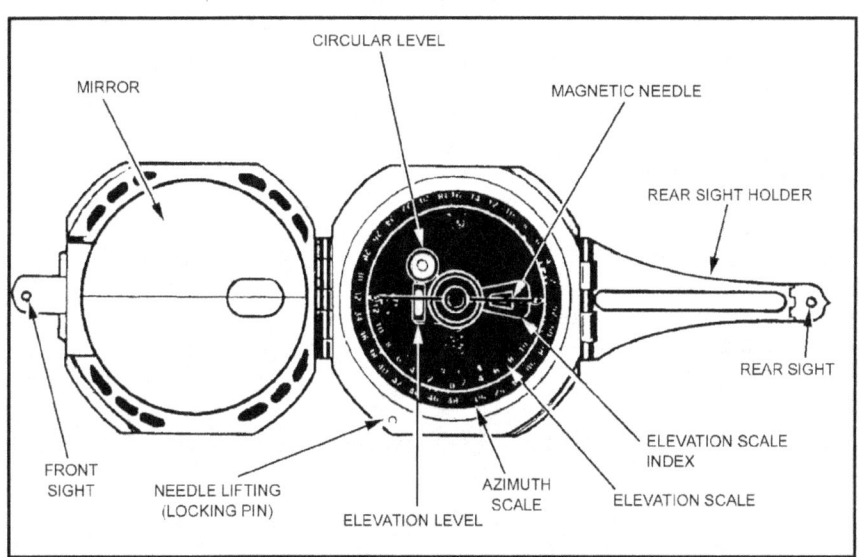

Figure F-1. Compass, M2, (top view).

F-1. CHARACTERISTICS
The main characteristics of the M2 compass are:
- Angle-of-site scale: 1200-0-1200 mils.
- Azimuth scale: 0 to 6400 mils.
- Dimensions closed: 2 3/4 inches by 1 1/8 inches.
- Weight: 8 ounces.

F-2. DESCRIPTION
The principal parts of the compass are described herein.

FOUO

a. **Compass Body Assembly.** This assembly consists of a nonmagnetic body and a circular glass window that covers the instrument, and keeps dust and moisture from its interior, protecting the compass needle and angle-of-site mechanism. A hinge assembly holds the compass cover in the position in which it is placed. A hole in the cover coincides with a small oval window in the mirror on the inside of the cover. A sighting line is etched across the face of the mirror.

b. **Angle-of-Site Mechanism.** The angle-of-site mechanism is attached to the bottom of the compass body. It consists of an actuating (leveling) lever located on the back of the compass, a leveling assembly with a tubular elevation level, and a circular level. The instrument is leveled with the circular level to read azimuths and with the elevation level to read angles of site. The elevation (angle-of-site) scale and the four points of the compass, represented by three letters and a star, are engraved on the inside bottom of the compass body. The elevation scale is graduated in two directions; in each direction it is graduated from 0 to 1200 mils in 20-mil increments and numbered every 200 mils.

c. **Magnetic Needle and Lifting Mechanism.** The magnetic needle assembly consists of a magnetized needle and a jewel housing that serves as a pivot. The north-seeking end of the needle is white. (The newer compasses have the north and south ends of the needle marked "N" and "S" in raised, white lettering.) On some compasses a thin piece of copper wire is wrapped around the needle for counterbalance. A lifting pin projects above the top rim of the compass body. The lower end of the pin engages the needle-lifting lever. When the cover is closed, the magnetic needle is automatically lifted from its pivot and held firmly against the window of the compass.

d. **Azimuth Scale and Adjuster.** The azimuth scale is a circular dial geared to the azimuth scale adjuster. This permits rotation of the azimuth scale about 900 mils in either direction. The azimuth index provides a means of orienting the azimuth scale at 0 or the declination constant of the locality. The azimuth scale is graduated from 0 to 6400 in 20-mil increments and numbered at 200-mil intervals.

e. **Front and Rear Sight.** The front sight is hinged to the compass cover. It can be folded across the compass body, and the cover closed. The rear sight is made in two parts—a rear sight and a holder. When the compass is not being used, the rear sight and holder are folded across the compass body and the cover is closed.

F-3. USE

The compass should be held as steadily as possible to obtain accurate readings. The use of a sitting or prone position, a rest for the hand or elbows, or a solid nonmetallic support helps eliminate unintentional movement of the instrument. When being used to measure azimuths, the compass must not be near metallic objects.

a. To measure a magnetic azimuth—

(1) Zero the azimuth scale by turning the scale adjuster.

(2) Place the cover at an angle of about 45 degrees to the face of the compass so that the scale reflection is viewed in the mirror.

(3) Adjust the front and rear sights to the desired position. Sight the compass by any of these methods:

(a) Raise the front sight and the extended rear sight assembly perpendicular to the face of the compass (Figure F-2 and Figure F-3, page F-4). Sight over the tips of the front and rear sights. If the object is above the line of sighting, fold the rear sight toward the eye as needed.

The instrument is correctly aligned when, with the level centered, the operator sees the tips of the sights and the center of the object at the same time.

(b) Raise the rear sight approximately perpendicular to the face of the compass. Sight on the object through the opening in the rear sight holder and through the window in the cover. Keep the compass level and raise or lower the eye along the opening in the rear sight holder until the black center line of the window bisects the object and the opening in the rear sight holder.

(c) Fold the rear sight holder out parallel with the face of the compass with the rear sight perpendicular to its holder. Sight through or over the rear sight and view the object through the window in the cover. If the object sighted is at a lower elevation than the compass, raise the rear sight holder as needed. The compass is correctly sighted when the compass is level and the operator sees the black center line of the window bisecting the rear sight and the object sighted.

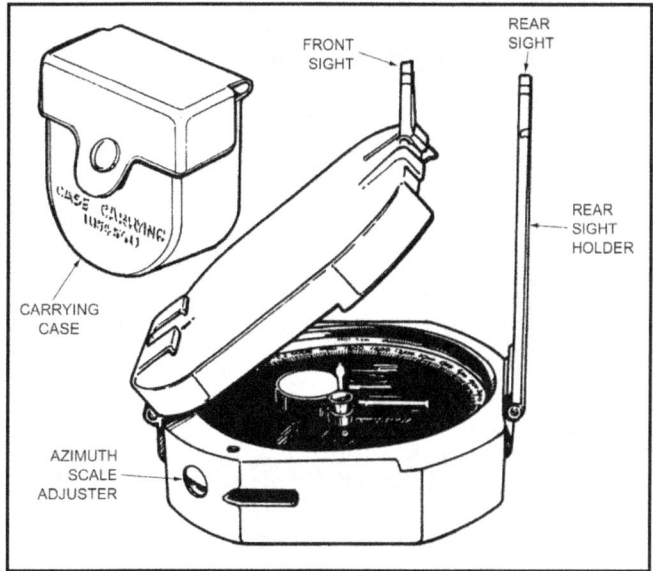

Figure F-2. Compass, M2 (side view).

FM 3-25.26

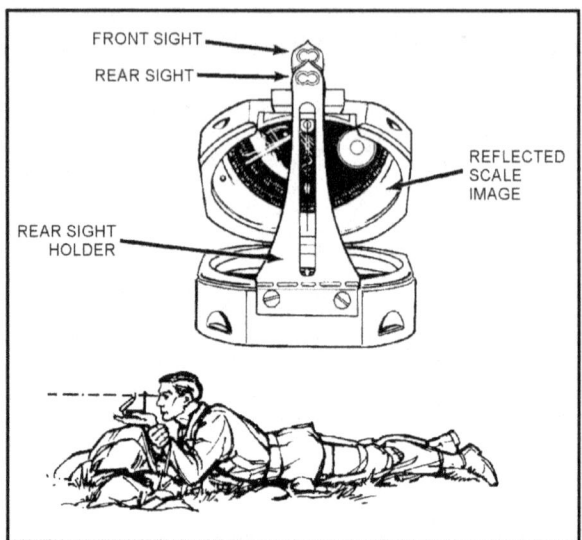

Figure F-3. Compass, M2 (user's view).

(4) Hold the compass in both hands, at eye level, with the arms braced against the body and the rear sight near the eyes. For precise measurements, rest the compass on a nonmetallic stake or object.

(5) Level the instrument by viewing the circular level in the mirror and moving the compass until the bubble is centered. Sight on the object, look in the mirror, and read the azimuth indicated by the black (south) end of the magnetic needle.

  b. To measure a grid azimuth—

(1) Index the known declination constant on the azimuth scale by turning the azimuth scale adjuster. Be sure to loosen the locking screw on the bottom of the compass. (The lightweight [plastic] M2 compass has no locking screw.)

(2) Measure the azimuth as described above. The azimuth measured is a grid azimuth.

  c. To measure an angle of site or vertical angle from the horizontal—

(1) Hold the compass with the left side down (cover to the left) and fold the rear sight holder out parallel to the face of the compass, with the rear sight perpendicular to the holder. Position the cover so that, when looking through the rear sight and the aperture in the cover, the elevation vial is reflected in the mirror.

(2) Sight on the point to be measured.

(3) Center the bubble in the elevation level vial (reflected in the mirror) with the level lever.

(4) Read the angle on the elevation scale opposite the index mark. The section of the scale graduated counterclockwise from 0 to 1200 mils measures plus angles of site. The section of the scale graduated clockwise from 0 to 1200 mils measures minus angles of site.

APPENDIX G
ADDITIONAL AIDS

This appendix provides information on the operation and function of already fielded, and soon to be fielded, devices that can be used as aids to navigation.

G-1. AN/PVS-5/5A, NIGHT VISION GOGGLES

These goggles are passive night vision devices. An infrared light source and positive control switch permit close-in viewing under limited illumination. The AN/PVS-5/5A has a field of view of 40 degrees and a range of 150 meters.

    a. The device has the capability for continuous passive operation over a 15-hour period without battery replacement. It weighs 1.5 pounds and is face-mounted. An eyepiece diopter is provided so the device can be worn without corrective lenses.

    b. The device is designed to assist the following tasks: command and control, fire control, reconnaissance, close-in surveillance, terrain navigation, first aid, operation and maintenance of vehicles, selection of positions, traffic control, rear and critical area security, patrolling, combat engineer tasks, radar team employment, resupply activities, and flight-line functions.

    c. It is a fielded system used by combat, CS, and CSS elements. The infantry, armor, air defense, field artillery, aviation, engineer, intelligence, military police, transportation, signal, quartermaster, chemical, maintenance, missile, and munitions units all use the device to help accomplish their missions.

    d. The AN/PVS-5/5A can assist the land navigator under limited visibility conditions. Chemical lights may be placed at selected intervals along the unit's route of movement, and they can be observed through the AN/PVS-5/5A. Another navigation technique is to have one person reading the map while another person reads the terrain, both using AN/PVS-5/5As. This allows the map reader and the terrain interpreter to exchange information on what terrain is observed, both on the map and on the ground. It allows each user to concentrate the AN/PVS-5/5A on one task. Land navigation, especially mounted, is a task better performed by more than one person. The above technique allows one soldier to perform map interpretation in the cargo portion of the vehicle while another soldier, possibly the driver, transmits to him information pertaining to the terrain observed on the ground.

G-2. AN/PVS-7B/D, NIGHT VISION GOGGLES

The AN/PVS-7B/D is a lightweight (1.5 pounds), image intensification, passive night-vision device that uses ambient light conditions. It has the same applications as the AN/PVS-5/5A. It is designed to be used in the same way as, and by the same units as, the AN/PVS-5/5A. The AN/PVS-7B/D has a field of view of 40 meters and a range of 300 meters in moonlight and 150 meters in starlight.

G-3. ENHANCED POSITION LOCATION REPORTING SYSTEM USER UNIT

The enhanced Position Location Reporting System (EPLRS)/Joint Tactical Information Distribution System (JTIDS) hybrid (PJH) is a computer-based system. It provides near real-time, secure data communications, identification, navigation, position location, and automatic reporting to support the need of commanders for information on the location, identification, and movement of friendly forces.

a. The EPLRS is based on synchronized radio transmissions in a network of users controlled by a master station. The major elements of an EPLRS community include the airborne, surface vehicular, and man-pack users; the EPLRS master station; and an alternate master station. The system can handle 370 user units in a division-size deployment per master station with a typical location accuracy at 15 meters. The man-pack unit weighs 23 pounds and includes the basic user unit, user readout, antenna, backpack, and two batteries.

b. The EPLRS is deployed at battalion and company level. Its use allows—

(1) Infantry or tank platoons to locate their positions, know the location of their friendly units, navigate to predetermined locations, and be informed when near or crossing boundaries.

(2) Artillery batteries to locate forward observers and friendly units, and position firing batteries.

(3) Aircraft to locate their exact positions; know the location of other friendly units; navigate to any friendly units, or a location entered by the pilot; navigate in selected flight corridors; and be alerted when entering or leaving corridors or boundaries.

(4) Command and control elements at all echelons to locate and control friendly units/aircraft.

c. The network control station is located at brigade level to provide position location/navigation and identification services. It also provides interface between the battalion and company systems, and the JTIDS terminals.

d. The EPLRS is fielded to infantry, armor, field artillery, military police, engineer, intelligence, aviation, signal, and air defense artillery units.

e. The EPLRS is a system that allows units to navigate from one point to another with the capability of locating itself and other friendly units equipped with the same system.

G-4. GLOBAL POSITIONING SYSTEM

The GPS is a space-based, radio-positioning navigation system that provides accurate passive position, speed, distance, and bearing of other locations to suitably equipped users.

a. The system assists the user in performing such missions as siting, surveying, tactical reconnaissance, sensor emplacement, artillery forward observing, close air support, general navigation, mechanized maneuver, engineer surveying, amphibious operations, signal intelligence operations, electronic warfare operations, and ground-based forward air control.

b. It can be operated in all weather, day or night, anywhere in the world; it may also be used during nuclear, biological, and chemical warfare.

c. It has been widely fielded in both active and reserve component units. (See Appendix I for more information on the GPS.)

G-5. POSITION AND AZIMUTH DETERMINING SYSTEM

The PADS is a highly mobile, self-contained, passive, all-weather, survey-accurate position/navigation instrument used by field artillery and air defense artillery units for fire support missions. Its basis of issue is two sets per artillery battalion. The device is about the size of a 3-kilowatt generator and weighs 322.8 pounds in operational configuration.

a. The two-man PADS survey party uses the HMMWV, the commercial utility cargo vehicle, the small-unit support vehicle, or the M151 1/4-ton utility truck. The system can be transferred while operating into the light observation helicopter (OH-58A) or driven into the CH-47 medium cargo helicopter.

b. The system provides real-time, three-dimensional coordinates in meters and a grid azimuth in mils. It also gives direction and altitude.

c. The PADS can be used by the land navigator to assist in giving accurate azimuth and distance between locations. A unit requiring accurate information as to its present location can also use PADS to get it. The PADS, if used properly, can assist many units in the performance of their mission.

> **WARNING**
> Laser devices are potentially dangerous. Their rays can and will burn someone's eyes if they look directly at them. Users should not direct the beams at friendly positions or where they could reflect off shiny surfaces into friendly positions. Other soldiers must know where lasers are being used and take care not to look directly at the laser beam.

G-6. GROUND-VEHICULAR LASER LOCATOR DESIGNATOR

The G/VLLD is the Army's long-range designator for precision-guided semi-active laser weapons. It is two-man portable for short distances and can be mounted on the M113A1 interim FIST vehicle when it has the vehicle adapter assembly. The G/VLLD provides accurate observer-to-target distance, vertical angle, and azimuth data to the operator. All three items of information are visible in the operator's eyepiece display.

a. The G/VLLD is equipped with an AN/TAS-4 nightsight. This nightsight increases the operator's ability to detect and engage targets during reduced visibility caused by darkness or battlefield obscuration.

b. The G/VLLD can give the navigator accurate line-of-sight distance to an object. The system can be used to determine its present location using resection and can assist the navigator in determining azimuth and distance to his objective.

G-7. QUICK RESPONSE MULTICOLOR PRINTER

The QRMP is a self-contained, laser, xerography printer capable of reproducing maps, photographs, annotated graphics, transparent originals, and digital terrain data in full color on transparent material or standard map paper. The QRMP system consists of a QRMP housed in an 8- by 8- by 20-foot ISO shelter mounted on a 5-ton truck with a dedicated military-standard 30-kilowatt generator. Each system will carry at least a seven-day supply of all necessary materials.

a. The QRMP system has map size (24- by 30-inch paper size and 22.5- by 29-inch image size), color printing, scanning and electronics subsystems. It produces the first copy in less than five minutes in full color and sustains a copy rate of 50 to 100 copies per hour for full color products. The system uses a charged couple device array for scanning and sophisticated electronic signal processing to electrostatically discharge a selenium photoreceptor drum.

b. The QRMP has the capability to print terrain and other graphics directly from digital output from the digital topographic support system or another QRMP.

APPENDIX H
FOREIGN MAPS

The use of foreign maps poses several problems to the land navigator. These products are often inferior in both content reliability and topographic accuracy to those produced by the DMA. Clues to these weaknesses are the apparent crudeness of the maps, unusually old compilation dates, or differences in mapped and actual terrain. The following characteristics should be examined closely.

H-1. HYDROGRAPHY
Of all the symbols on foreign maps, those for hydrography conform most closely to NGA usage. The use of blue lines and areas to depict streams, rivers, lakes, and seas seems to be universally accepted. The one caution to be observed is that foreign cartographers use different sets of rules to govern what is and what is not included on the map. Distinction between perennial and intermittent streams is usually not made.

H-2. VEGETATION
The classification and symbols for vegetation on most foreign maps are different to those used on NGA maps. The vegetation included on many foreign maps is often extensive, identifying not only vegetated areas, but also the specific types of vegetation present. Green is the predominant color used to represent vegetation although blue and black are sometimes used. The symbols that depict the various types of vegetation differ greatly from one foreign map to another.

H-3. CULTURAL AND LINEAR FEATURES
Perhaps the most striking difference between NGA and foreign maps is the set of symbols used to portray cultural features. Some symbols found on foreign maps are very unusual. Symbols for linear features on foreign maps are also likely to confuse the user who is accustomed to NGA symbols. NGA uses ten basic road symbols to portray different classes of roads and trails; foreign cartographers use many more.

H-4. TERRAIN RELIEF
Foreign maps generally use contour lines to portray terrain relief, but substantial variability exists in the contour intervals employed. They may range from 5 to 100 meters.

H-5. SCALE
Scales found on foreign maps include 1:25,000, 1:63,360, 1:63,600, 1:75,000, and 1:100,000. Most foreign large-scale topographic maps have been overprinted with 1,000-meter grid squares so it is unlikely that the variable scales will have much effect on your ability to use them. However, you must learn to estimate grid coordinates because your 1:25,000 and 1:50,000 grid coordinate scales may not work.

H-6. STEPS TO INTERPRETING FOREIGN MAPS

After discussing the many difficulties and limited advantages encountered when using foreign maps, it is only appropriate that some strategy be offered to help you with the task.

a. In the August 1942 issue of The Military Engineer, Robert B. Rigg, Lieutenant, Cavalry, suggested a five-step process for reading and interpreting foreign maps. It is as appropriate today as it was when he first proposed it.

Step 1. Look for the date of the map first. There are generally four dates: survey and compilation, publication, printing and reprinting, and revision. The date of the survey and compilation is most important. A conspicuous date of revision generally means that the entire map was not redrawn—only spot revisions were made.

Step 2. Note whether the publisher is military, government, or civilian. Maps published by the government or the military are generally most accurate.

Step 3. Look at the composition. To a great extent, this will reveal the map's accuracy. Was care taken in the cartography? Are symbols and labels properly placed? Is the draftsmanship precise? Is the coastline or river bank detailed?

Step 4. Observe the map's color. Does it enhance your understanding or does it obscure and confuse? The importance of one subject (coloring) must warrant canceling others. If it confuses, the map is probably not very accurate.

Step 5. Begin to decode the various map colors, symbols, and terms. Study these items by examining one feature classification at a time (culture, hydrography, topography, and vegetation). As an accomplished navigator, you should already have a good understanding of your area of operations, so translation of the map's symbols should not present an impossible task. Use your notebook to develop an English version of the legend or create a new legend of your own.

b. In dealing with the challenge of using a foreign map, be certain to use these five steps. In doing so, you are also encouraged to bring to bear all that you know about the geographic area and your skills in terrain analysis, map reading, map interpretation, and problem solving. After careful and confident analysis, you will find that what you do know about the foreign map is more than what you do not know about it. The secret often lies in the fact that the world portrayed on a map represents a kind of international language of its own, which allows you to easily determine the map's accuracy and to decode its colors, symbols, and labels.

APPENDIX I
GLOBAL POSITIONING SYSTEM

The ability to accurately determine position location has always been a major problem for soldiers. However, the global positioning system has solved that problem. Soldiers will now be able to determine their position accurately to within 10 meters.

I-1. DEFINITION
The GPS is a satellite-based, radio navigational system. It consists of a constellation with 24 active satellites that interfaces with a ground-, air-, or sea-based receiver. Each satellite transmits data that enables the GPS receiver to provide precise position and time to the user. The GPS receivers come in several configurations; hand-held, vehicular-mounted, aircraft-mounted, and watercraft-mounted.

I-2. OPERATION
The GPS is based on satellite ranging. It figures the user's position on earth by measuring the distance from a group of satellites in space to the user's location. For accurate three-dimensional data, the receiver must track four or more satellites. Most GPS receivers provide the user with the number of satellites that it is tracking, and whether or not the signals are good. Some receivers can be manually switched to track only three satellites if the user knows his altitude. This method provides the user with accurate data much faster than that provided by tracking four or more satellites. Each type receiver has a number of mode keys that have a variety of functions. To better understand how the GPS receiver operates, refer to the operator's manual.

I-3. CAPABILITIES
The GPS provides worldwide, 24-hour, all-weather, day or night coverage when the satellite constellation is complete. The GPS can locate the position of the user accurately to within 21 meters—95 percent of the time. However, the GPS has been known to accurately locate the position of the user within 8 to 10 meters. It can determine the distance and direction from the user to a programmed location or the distance between two programmed locations called way points. It provides exact date and time for the time zone in which the user is located. The data supplied by the GPS is helpful in performing several techniques, procedures, and missions that require soldiers to know their exact location. Some examples are:
- Sighting.
- Surveying.
- Sensor or minefield emplacement.
- Forward observing.
- Close air support.
- Route planning and execution.
- Amphibious operations.
- Artillery and mortar emplacement.
- Fire support planning.

I-4. LIMITATIONS

A constellation of 24 satellites broadcasts precise signals for use by navigational sets. The satellites are arranged in six rings that orbit the earth twice each day. The GPS navigational signals are similar to light rays, so anything that blocks the light will reduce or block the effectiveness of the signals. The more unobstructed the view of the sky, the better the system performs.

I-5. COMPATABILITY

All GPS receivers have primarily the same function, but the input and control keys vary between the different receivers. The GPS can reference and format position coordinates in any of the following systems:

- **Degrees, Minutes, Seconds (DMS):** Latitude/longitude-based system with position expressed in degrees, minutes, and seconds.
- **Degrees, Minutes (DM):** Latitude/longitude-based system with position expressed in degrees and minutes.
- **Universal Traverse Mercator (UTM):** Grid zone system with the northing and easting position expressed in meters.
- **Military Grid Reference System (MGRS):** Grid zone/grid square system with coordinates of position expressed in meters.

The following is a list of land navigation subjects from other sections of this manual in which GPS can be used to assist soldiers in navigating and map reading:

   a. **Grid Coordinates (Chapter 4).** GPS makes determining a 4-, 6-, 8-, and 10-digit grid coordinate of a location easy. On most GPS receivers, the position mode will give the user a 10-digit grid coordinate to their present location.

   b. **Distance (Chapter 5) and Direction (Chapter 6).** The mode for determining distance and direction depends on the GPS receiver being used. One thing the different types of receivers have in common is that to determine direction and distance, the user must enter at least one way point (WPT). When the receiver measures direction and distance from the present location or from way point to way point, the distance is measured in straight line only. Distance can be measured in miles, yards, feet, kilometers, meters, or nautical knots or feet. For determining direction, the user can select degrees, mils, or rads. Depending on the receiver, the user can select true north, magnetic north, or grid north.

   c. **Navigational Equipment and Methods (Chapter 9).** Unlike the compass, the GPS receiver when set on navigation mode (NAV) will guide the user to a selected way point by actually telling the user how far left or right the user has drifted from the desired azimuth. With this option, the user can take the most expeditious route possible, moving around an obstacle or area without replotting and reorienting.

   d. **Mounted Land Navigation (Chapter 12).** While in the NAV mode, the user can navigate to a way point using steering and distance, and the receiver will tell the user how far he has yet to travel, and at the current speed, how long it will take to get to the way point.

   e. **Navigation in Different Types of Terrain (Chapter 13).** The GPS is capable of being used in any terrain, especially more open terrain like the desert.

   f. **Unit Sustainment (Chapter 14).** The GPS can be used to read coordinates to quickly and accurately establish and verify land navigation courses.

APPENDIX J
PRECISION LIGHTWEIGHT GPS RECEIVER

The precision lightweight GPS receiver (PLGR) is a highly accurate satellite signal navigation set (referred to in this appendix as AN/PSN-11).

J-1. CONCEPT OF OPERATION

The AN/PSN-11 is designed for battlefield use anywhere in the world. It is sealed watertight for all-weather day or night operation. The AN/PSN-11 is held in the left hand and operated with the thumb of the left hand. Capability is included for installation in ground facilities, and air, sea, and land vehicles. The AN/PSN-11 is operated stand-alone using prime battery power and integral antenna. It can be used with external power source and external antenna.

   a. The AN/PSN-11 provides the user with position coordinates, time, and navigation information under all conditions, if—
   - No obstructions block the line-of-sight satellite signal from reaching the antenna.
   - Valid crypto keys are used to protect the AN/PSN-11 from intentionally degraded satellite signals.

   b. Many data fields, such as elevation, display units of information. The format of the units can be changed to your most familiar format.

   c. Map coordinates are entered as a waypoint. When a waypoint is selected as a destination, the AN/PSN-11 provides steering indications, azimuth, and range information to the destination. A desired course to a waypoint is entered. Offset distance from this course line is shown.

   d. Up to 999 waypoints can be entered, stored, and selected as a destination. A route is defined for navigation either start-to-end or end-to-start. The route consists of up to nine legs (ten waypoints) linked together.

J-2. CAPABILITIES

Data provided by the AN/PVS-11 helps complete missions such as:
- Siting.
- Surveying.
- Tactical reconnaissance.
- Sensor emplacement.
- Artillery forward observing.
- Close air support.
- General navigation.
- Mechanized maneuvers.
- Engineer surveying.
- Amphibious operations.
- Parachute operations.
- Signal intelligence.

FM 3-25.26

- Electronic warfare.
- Ground-based forward air control.

This data is displayed on the AN/PSN-11display. It is also available from a serial data port.

J-3. CHARACTERISTICS

The AN/PSN-11 is less than 9.5 inches long, 4.1 inches wide, and 2.6 inches deep. It weighs 2.75 pounds with all batteries in place. The small size and light weight make the set easy to carry and use. The durable plastic case is sealed for all-weather use. The AN/PSN-11 features make it easy to use. (These features are highlighted in the physical description in Figure J-1).

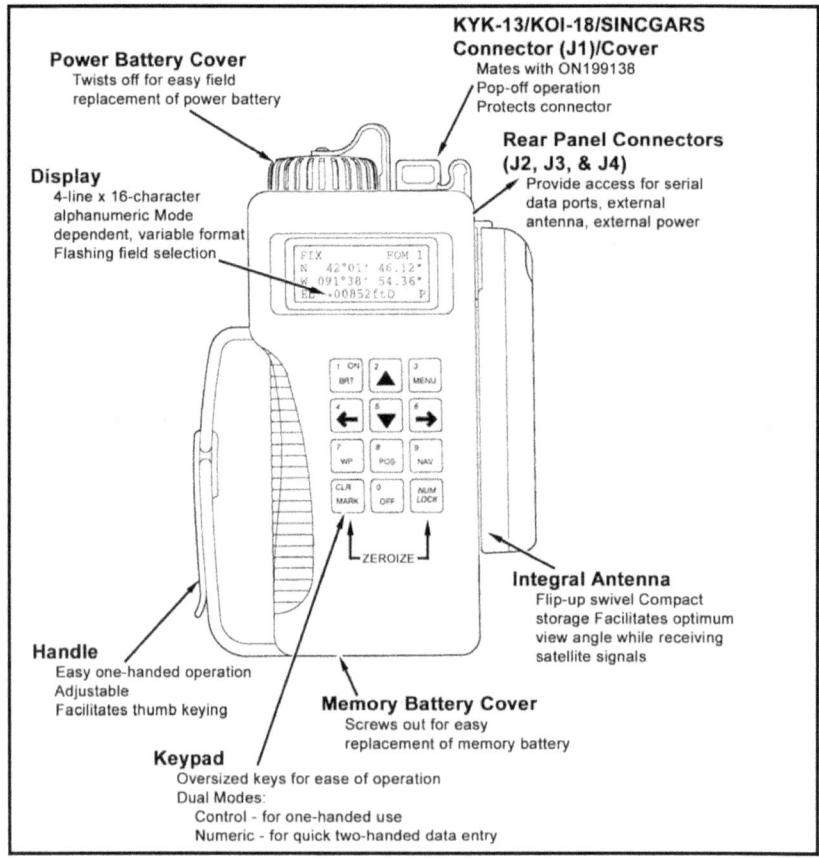

Figure J-1. Physical features of the AN/PSN-11.

J-4. SETUP AND CONTROL

Setting up the operation parameters of the AN/PSN-11 is critical. This section describes the display and the procedures and principles used in setting the display to suit the needs of the user. This display consists of seven pages that allows the user to control the following parameters:

- Operating mode.
- Type of satellites to use.
- Coordinate system.
- Units.
- Magnetic variation.
- Display customization.
- Navigation Display mode.
- Elevation hold mode.
- Time and error formats.
- Datum.
- Automatic off timer.
- Datum port configuration.
- AutoMark mode.

Perform the following procedures to set up the AN/PSN-11 for continuous operation:

a. Turn the AN/PSN-11 **ON**. Once it has completed its built-in-test (BIT) press the **MENU** key and move the cursor to **SETUP** (Figure J-2). Activate the SETUP function.

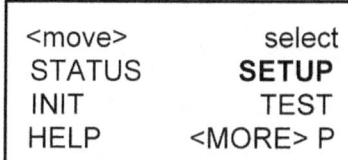

Figure J-2. SETUP.

b. The first screen (Figure J-3) allows the operator to set the operating mode and SV-type. Scroll through the operating modes and select **CONT** and for the SV-type **Mixed**.

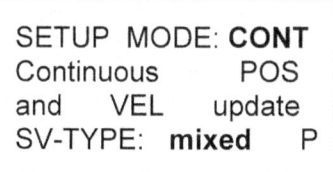

Figure J-3. Operating mode and SV-type.

c. The second screen (Figure J-4) allows the operator to set up the units. Scroll through the available coordinates and select **MGRS-New** and **Metric**. For the Elevation select **meter** and **MSL** and for the Angle select **degrees** and **magnetic**.

```
SETUP            UNITS
MGRS-New         Metric
Elev: meter      MSL
ANGL: Deg    Mag     P
```

Figure J-4. Set up the units.

d. The third screen (Figure J-5) should be set for the MAGVAR (magnetic variation or GM angle for your area). The operator can select "calculate the degree" or manually enter degrees as an Easterly or Westerly GM angle; for example, **E021.0** for the TENINO map sheet.

```
SETUP      MAGVAR
TYPE: Calc    deg
         WWM    1995
                   P
```

Figure J-5. Magnetic variation or GM angle setup.

e. The fourth screen (Figure J-6) of setup allows the operator to set the elevation hold, time, and error. The operator should set the ELHOLD to **automatic**. As for time, the operator needs to know, from their present location, how many hours they are ahead of or behind Greenwich mean time. For example, during daylight savings time, Fort Benning, GA. is **Loc=Z-0400**. To set the ERR, the operator selects **-+m** to let him know in meters how accurate the PLGR is operating.

```
SETUP
ELHold :    automatic
TIME:       Loc=Z-0400
ERR: +-m              P
```

Figure J-6. Set elevation, time, hold, and error.

f. The fifth screen (Figure J-7) of setup allows the operator to set the AN/PSN-11 datum to his area of operation and to set the automatic off timer. The AN/PSN-11 has fifty-two map datum sets available. The operator should set the datum to his area of operation. For example, if your map datum is WGS-84, the operator sets the AN/PSN-11 to WGS-84. If the map is 1927 North America datum, the operator sets the datum to NAS-C. The automatic timer off is used to turn the AN/PSN-11 off after a prescribed time once it has acquired a fixed position. The operator should set this mode to **OFF**.

```
SETUP DTM:NAS-C
NA27CONUS  /Clk66
AUTOMATIC OFF
TIMER: off          P
```

Figure J-7. Set the PLGR datum.

g. The sixth screen (Figure J-8) in setup is the in/out port screen. This page allows the operator to control serial communications, HAVEQUICK and 1PPS options. Select **Standard** unless otherwise directed and select **OFF** for HAVEQUICK and 1PPS.

```
SETUP I/O
SERIAL: Standard
HAVEQUICK: Off
1PPS: Off           P
```

Figure J-8. In/out port screen.

h. The seventh screen (Figure J-9) is setup AUTOMARK. This feature allows the operator to have the AN/PSN-11 periodically wake up, acquire a position fix, store the position as a waypoint, or return to the mode of operation it was previously in. The operator should set this mode to **OFF**. The remaining pages for SETUP are for advanced GPS users.

```
SETUP    AUTOMARK
MODE: off    WP002
26-04-01        0935L
REPEAT 00h00m    P
```

Figure J-9. AUTOMARK setup.

i. Once the AN/PSN-11 is set up, the operator can obtain a position. This procedure is accomplished by activating the position (POS) key. The position displayed is **old** information until the receiver collects and calculates satellite data and displays the current position. The receiver must be tracking three satellites to obtain a two-dimensional fix position and four or more satellites for a three-dimensional fix position. The third dimension is elevation.

J-5. WAYPOINT OPERATIONS

A waypoint is the location of a point on a desired course described by coordinates or a physical location. A normal mission consists of a series of waypoints. The waypoints available on the AN/PSN-11 are 999 (numbered 01 through 999).

a. This paragraph describes the AN/PSN-11 waypoint displays and waypoint operations. The waypoint display pages are used to perform the following operations:

- Enter, edit, or review waypoints.
- Copy waypoints.
- Determine the distance between waypoints.
- Calculate a new waypoint.
- Clear waypoints.
- Define a mission route.

b. To enter a waypoint, the operator needs to press the waypoint (**WP**) key (Figure J-10). When the waypoint menu appears, the ENTER function flashes. The operator presses the down arrow key to activate this field. Now the operator enters a waypoint name, grid zone designator, 100,000-meter grid square identifier, 10-digit grid coordinate, and elevation.

```
WP     <move>    sel
ENTER EDIT COPY
SR-CALC RNG CALC
DIST CLEAR ROUTE
```

Figure J-10. Enter a waypoint.

c. To enter a waypoint name, the operator presses the right arrow key until the first letter of the word UNUSED(WP#) is flashing (Figure J-11). Scroll up or down through the alphabet changing the letter U to whatever is desired. For example, if the operator wanted to name their waypoint NORTH STAR, the operator scrolls down the alphabet until the letter U is changed to the letter N (Figure J-12). The operator repeats this process for the remaining letters.

```
WP002 UNUSED002
B          MGRS-New
AN  00000e  00000n
No EL       CLR  P
```

Figure J-11. Unused.

```
WP002 NORTHSTAR
10T        MGRS-New
EG  13130e   95750n
No EL          CLR N
```

Figure J-12. Change a name.

d. Second line, the operator enters the grid zone designator for his area of operation. For example, the Fort Benning area falls in the 16S zone.

e. Third line, the operator must enter a 10-digit grid coordinate with its 100,000-meter grid square identifier. For example, if the waypoint location is Offutt Lake, Tenino map sheet, the 100,000-meter grid square identifier is EG. Then, the operator plots the grid coordinates on the map and enters it into the AN/PSN-11.

NOTE: Operator plots 8-digit grid coordinates, however a 10-digit coordinate is entered. Therefore, the 5th and 10th digit entered is a zero (0).

f. For the fourth line, if the elevation of the waypoint is known, the operator can enter it. If the elevation is not known the operator can just leave the data as zero or No EL. The operator moves the cursor until the up and down arrow symbol appears before the letter P or N in the bottom right corner. When activating the down arrow key the operator stores the waypoint into the AN/PSN-11's memory. The AN/PSN-11 notifies the operator that the waypoint has been stored.

NOTE: When entering numbers, the NUM LOCK can be activated. The letter N appears in the bottom right corner allowing the operator to use the numbers on the keypad rather then scrolling up or down.

J-6. NAVIGATION

Navigation (nav) is using the AN/PSN-11 to find your present position, relative to other points. The AN/PSN-11 provides azimuth, range, and steering information in a variety of formats. There are four navigation display modes that may be accessed and selected. The navigation display mode selected determines the type of information shown on the navigation displays. These navigation displays give the user the most useful information for a certain mission profile: **SLOW, 2D FAST, 3D FAST OR CUSTOM.**
- In **SLOW** navigation mode, the AN/PSN-11 performs two-dimensional (2D) navigation. SLOW navigation mode is used for land or sea navigation, when the user cannot maintain the minimum speed necessary (about 1.5 kilometers per hour).
- In **2D FAST** navigation mode, the AN/PSN-11 performs two-dimensional (2D) navigation. 2D FAST navigation mode is used for land or sea navigation, when the user can maintain the minimum speed necessary for GPS to compute navigation parameters based on velocity.

- In **3D FAST** navigation mode, the AN/PSN-11 performs three-dimentional (3D) navigation. 3D FAST navigation mode has an APPROACH sub-mode. 3D FAST navigation mode is used for air navigation, when the user can travel in three dimensions and can maintain the minimum speed necessary for GPS to compute navigation parameters based on velocity.
- In **CUSTOM** navigation mode, the AN/PSN-11 performs the user's navigational display pages as so desired. It can be set up to support the individual user's performances or mission requirements. The custom display modes available are Direct, Course To, Course From, Route, and Approach.

To navigate on land using a dead-reckoning method, the AN/PSN-11 navigation mode is accomplished as follows.

a. The operator presses the NAV key activating the navigation function. The first screen that appears is the navigation mode (Figure J-13). For example, **SLOW, 2D FAST, 3D FAST, CUSTOM, DIRECT, CRS TO, and CRS FROM.**

```
2D   FAST         DIRECT
WP002 NORTHSTAR002

                         P
```

Figure J-13. Navigation mode.

b. The operator selects the **2D FAST** and **DIRECT**. The second line is the waypoint to be navigated. (Scroll through the waypoints that are stored to choose the desired waypoint.)

c. To see the azimuth that the navigator should be traveling on, go to the next page by pressing the down arrow key (Figure J-14). This page tells the navigator what azimuth they are heading on (TRK=tracking), and the actual azimuth the navigator should be heading on (AZ). The fourth line tells the navigator steering (STR), a direction (< >), and a number of degrees the navigator needs to move to travel on the actual azimuth.

Figure J-14. Azimuth.

d. The third screen (Figure J-15) tells the navigator the range or distance to their waypoint and how much time (TTG2) it will take them to get to their waypoint. This page also lets the navigator know what the elevation difference is from their present location to the waypoint and by how much they will miss their waypoint by (MMD).

```
RNG   3598.55km
TTG2    0036:05
ELD    -00050m
MMD2  30m                P
```

Figure J-15. Range or distance.

This Page intentionally left blank.

APPENDIX K
DEFENSE ADVANCED GPS RECEIVER

The defense advanced GPS receiver (DAGR) is a handheld or host platform-mounted device that receives and decodes radio frequency (RF) signals from GPS satellite Link One (L1) and Link Two (L2). It provides position, velocity (ground speed), and time (PVT) reporting, and navigation capabilities. Although the DAGR has many features, this appendix only covers the procedures to place the unit into operation, create waypoints, navigate with the DAGR, use the situational awareness function, and troubleshoot.

K-1. INTRODUCTION

The DAGR's primary function is to navigate through terrain using stored waypoint position information. The DAGR is also used in operations such as waypoint calculations, data transfer, targeting, determining jamming sources, gun laying, and man overboard.

a. The DAGR is primarily a handheld unit with a built-in integral antenna, but can be installed in a host platform (ground facilities; air, sea, and land vehicles) using an external power source and an external antenna (Figure K-1, page K-2). The DAGR used as a handheld unit can also operate with an external L1/L2 antenna and a source of external power. Table K-1 describes some of its characteristics and capabilities.

SELECTED CHARACTERISTICS AND CAPABILITIES	
Length:	6.35 inches
Width:	3.46 inches
Depth:	1.58 inches
Weight w/batteries:	1 pound
Number of waypoints in memory:	999
Number of routes in memory:	15 with up to 1,000 legs for each
Miscellaneous capabilities:	<ul><li>Provides signal acquisition using up to 12 channels.</li><li>All satellites in view are tracked using 11 channels.</li><li>Provides navigation using up to 10 channels.</li><li>Produces no signals that can reveal your position.</li><li>Resists jamming.</li><li>Determines and stores the azimuth of a jamming signal source.</li><li>Uses an internal compass to compute track and ground speed when moving at or below 0.5 meter per second.</li><li>Compatible with NVGs and does not cause blooming.</li><li>Maps can be loaded.</li></ul>

Table K-1. DAGR characteristics and capabilities.

FM 3-25.26

Figure K-1. Defense advanced GPS receiver (DAGR).

b. Crypto variable (CV) keys may be loaded into the DAGR for increased PVT accuracy and protection from intentional false or spoofed satellite signals.

c. Mission data can be selectively cleared or zeroized at any time

K-2. CONTROLS AND INDICATORS

The DAGR has functions keys located beneath the display, which contains three display windows. The operator accesses the various DAGR functions by bringing up and selecting items from menus. Data is changed by using the cursor keys.

a. **Keypad Controls** (Figure K-2). The DAGR control keys can be used to perform two actions. The operator can push and hold the key to access one function or push and release the key to access another. Table K-1 describes the keys and their associated functions.

Figure K-2. Function keys.

K-2 FOUO
18 January 2005

Key	Push and Hold	Push and Release
F1/IN Function Key	F1 functions*	Zooms in on the situational awareness and map pages.
F2/OUT Function Key	F2 functions*	Zooms out on the situational awareness and map pages.
F3/STATUS Function Key	F3 functions*	Displays the current DAGR status.
PWR/QUIT Key	Turns the DAGR off or on.	• Cancels an operation. • Pages backwards when using a page set. • Returns to a previous display in a series of operational displays.
POS/PAGE KEY	Present position page	Scrolls to the next page of data.
BRIGHTNESS/MENU Key	Toggles the keypad and turns display lighting on and off.	Accesses display menus.
WP/ENTER Key	Selects between different waypoint functions.	• Selects items from pop-up menus. • Selects a field (highlight) when no field is currently selected. • Makes choices within lists.
* The functions these keys access are displayed sequentially from left to right on the toolbar at the bottom of the display.		

Table K-2. DAGR function keys.

b. **Editing Data Within the Display Window.** After the steps are completed to edit a field, the DAGR uses the same process for all operations to select and edit data in the display window. Use the cursor control keys to make selections or enter data in the display window. The left, right, up, and down cursor control keys function as follows:

(1) Push and release a cursor control key for one scroll (movement) of the cursor from field to field or option to option in the display.

(2) Push and hold a cursor control key for an accelerated scroll in the desired direction.

(3) Up and down cursor control keys are used to scroll through data vertically within a selected field as well as to move from field to field. For example, to enter new waypoint coordinates move to each digit, make the correction, push ENTER to confirm the change, and then move to the next digit.

(4) The left and right cursor control keys are used to scroll through data horizontally as well as move from field to field.

(5) The procedure to select and enter data into fields depends on how the data is presented and its type.

(a) *Editing Field Options from a List.* A list editor is a pop-up containing a menu of choices not requiring any individual character editing. Scroll to the correct entry and select it by pushing ENTER.

(b) *Editing Fields Containing Only Numbers.* A number editor is a pop-up containing numeric characters for editing. Use the left and right cursor keys to select the digit to edit. Use the up and down cursor keys to select the new digit, then push ENTER.

(c) *Editing Fields Containing Alphanumeric Characters.* A text editor is a pop-up containing alphanumeric characters for editing. Use the cursor control keys to select the character to edit, then push ENTER.

c. **Multifunction Keys.** The DAGR also has two sets of multifunction keys.

(1) *PWR/QUIT and POS/PAGE Key.* Push and release the PWR/QUIT and POS/PAGE keys simultaneously to activate the emergency zeroize display. Confirmation from the user is required before the action is completed.

(2) *BRIGHTNESS/MENU Key and Up or Down Cursor Control Key* When the keypad/display lighting is on, push and hold the BRIGHTNESS/MENU key and push the respective up or down cursor control key simultaneously to adjust lighting brightness level.

d. **Lighting, Battery, and Function Indicators and Labels.** The DAGR displays the lighting status, the battery strength, and the function being used in the display window.

(1) *Lighting Status Indicator.* The lighting status indicator is located in the upper right corner of the display next to the battery status indicator. It resembles a light bulb when the keypad/display lighting is on, and it does not appear when the lighting is off.

(2) *Battery Status Indicator.* The battery status indicator is located in the upper right corner of the display. It resembles a battery and the darkened portion indicates how much battery life remains.

(3) *Function Key Labels.* Each of the three function keys of the keypad has an associated function key label shown in the toolbar window at the bottom of the display. The function key actions are activated by pushing and holding the respective key on the keypad directly below the toolbar window.

e. **Display Windows** (Figure K-3). The DAGR display contains three windows: page, toolbar, and message. The page and toolbar windows are also divided into two sections that are always visible. The message window appears as needed to display additional messages, including pop-up information.

(1) The display windows cannot be individually selected; only fields included in the windows may be selected. The operator can request help text or a menu specific to the currently displayed page when no fields are selected.

(2) The page window is where the majority of display interaction occurs. A page may contain several individual fields. The fields may contain "read only" data or data that can be modified. A page may contain multiple horizontal or vertical views, as denoted by scrollbars at the right side or bottom of the page window. The scrollbars are controlled by the cursor control keys.

Figure K-3. DAGR display.

(3) The toolbar window consists of three display sections and is located at the bottom of the screen. It displays labels for the push and hold keys referred to as function keys (F1, F2, and F3). The function keys are used to change the page being displayed or to perform a single action (for example, go to the NAV display page).

f. **Message.** When conditions warrant operator notification, message windows are used to attract the operator's attention. Messages are categorized as notes, alerts, cautions, and warnings based on the impact of the message to the operator's mission. The message window is displayed over the page window. The message must be cleared (through operator acknowledgement or self-removal) before the page window functionality can be resumed.

g. **Pop-Ups.** Message, menu, help, and editor pop-ups are displayed over the page window. The operator initiates a pop-up by pushing the MENU key, or by pushing the ENTER key when a field is selected. The pop-up is cleared by making a selection from the pop-up display or pushing the QUIT key. Page window functionality is resumed after removing the pop-up. Pop-ups may have menus, allow editing, and have help text pop-ups associated with the displayed information.

h. **Editors.** The DAGR provides a variety of editors for the operator to change or customize page field content. Editors are accessed through the page or field menu. Actual DAGR editor titles correspond with the field being edited (for example, when editing a waypoint name field, the text editor title is Name). The operator primarily uses the number editor when selecting courses and creating waypoints. (See TM 11-5820-1172-13 for more details on how to use the DAGR edit features.)

(1) *Number Editor.* The number editor is used when editing numeric field values (for example, grid coordinates). The number editor utilizes key functions as follows:
- Up/down cursor control keys — scroll to desired digit or characters.
- PAGE key — scroll to the first digit or character value.
- Left/right cursor control keys — move the cursor left or right (for example, move to the next digit in a coordinate).
- ENTER key — save changes and exit.
- QUIT key — exit without saving changes

(2) *List Editors* (Figure K-4). The list editor utilizes key functions the same as the number editor except the PAGE key is used to scroll down larger lists. The list editor is used when editing operator selectable data (for example, selecting from a pick list). List editors are also used for special lists (for example, including both waypoint number and name) or additional information of the highlighted item in a display footer (for example, datum information).

Figure K-4. List editors.

(3) *Text Editors.* The text editor is used when editing text and numeric characters. It allows selection of the characters A through Z, 0 through 9, dash (-), slash (/), period (.), and space ( ) to be entered into the text box. To select/activate a given key, the ENTER key must be used. There are four command keys: Clear, Ins Char, Del Char, and Save.

(a) *Clear.* The Clear key clears the selected character and all characters to the right of the selected character and replaces them with the space character.

(b) *Ins Char.* Insert Character shifts the selected character and all characters following the selected character to the right by one character, and inserts a space character at the selected location. The new space character becomes the selected character and the character at the end of the text string is deleted (last character of the last line of editable text).

(c) *Del Char.* Delete Character shifts all characters following the selected character to the left by one character, thereby overwriting the selected character, and a space character is inserted at the end of the text string (last character of the last line of editable text). The character replacing the selected character becomes the new selected character.

(d) *Save.* Saves the changes made to the text string and exits the text editor. Instead of using the cursor keys to highlight the SAVE command, the MENU key can be pushed to access a list of options (undo changes, save and exit, exit and no save, reset to default, and editor help). These options provide shortcuts to close the text editor.

i. **Menus.** The DAGR uses the following general menu structure to access and or edit information. The four menu types are:
- Main menu: provides submenu choices.
- Submenu: provides page (function) choices.

- Page menu: provides specific functions or editors associated with the page.
- Field menu: provides specific functions or editors associated with the field.

(1) With a page displayed or a field highlighted, the corresponding menu may be viewed by pushing the MENU key. Pushing the QUIT key allows the user to back out of the menu and return to the previous display. When a highlighted menu selection has an arrow symbol to its right, pushing the right cursor control key or the ENTER key causes the submenu to be displayed.

(2) Field and page menu items that are not currently available (for example, Edit Field) appear as light gray text. Although the cursor can be placed on disabled items, the pop-up menu does not allow selection of that item.

j. **Accessing Menus.** DAGR menus are accessed in steps. The operator must access each level, select the next menu, and proceed down the hierarchy until reaching the desired field. The operator uses the cursor keys to highlight and the ENTER key to select.

(1) *Going from the Main Menu to a Field Menu.* Use the steps to access the desired field menu.

(a) Access the main menu and select the correct submenu by highlighting it and pushing ENTER.

(b) From the submenu, select the correct page menu by highlighting it and pushing ENTER.

(c) From the page menu, select the correct field menu by highlighting it and pushing ENTER.

(2) *Accessing the Main Menu.* Following DAGR power-up, and from any display (except a pop-up message), access the main menu by pushing the MENU key twice. With a submenu open, the main menu is accessed by pushing the QUIT key. When a page menu or a field menu is open, the main menu is accessed by pushing the MENU key.

(3) *Accessing the Waypoint Page.* The operator can immediately access the waypoint functions by pushing and holding the WP/ENTER key.

NOTE: When a page is displayed, other pages of the submenu page set are accessed by pushing the PAGE or QUIT keys.

k. **POS/PAGE Set.** POS/PAGE set speeds access to the most commonly used DAGR functions. It is accessed by pushing and holding the POS/PAGE key. After accessing the POS/PAGE set, the PAGE or QUIT key can be pushed to view all pages of the POS/PAGE set. The POS/PAGE set consists of the following pages:

- Present Position.
- Situational Awareness.
- NAV Pointer.
- Map.
- SV Sky View.

NOTE: The operator can remove the SV Sky View and Map page from the POS/PAGE set, but not the Present Position, Situational Awareness, and NAV Pointer pages. The operator can add up to seven additional display pages to the nonremovable pages of the POS/PAGE set for a total of up to ten display pages.

(1) ***Present Position*** (Figure K-5). The Present Position page displays the operator's present position and contains ten fields. The operator can scroll the page to view the additional field data. The ten fields include: Present Position Coordinates, Coordinate and Grid System, Datum Identifier, Current Operating Mode, EHE, FOM, Elevation Reference, Ground Speed, Estimate Time Error, and MAGVAR (magnetic variance)

Figure K-5. Present position.

(2) ***Situational Awareness.*** The Situational Awareness page provides a graphical display of relationships between present position, track, waypoints, routes, and alerts. The Situational Awareness page includes a north reference indicator, speed and track, position error data, and a range scale.

(3) ***NAV Pointer*** (Figure K-6). The NAV Pointer page displays a pointer directing the operator towards the displayed waypoint. It also displays the current navigation method, destination waypoint number and name, azimuth, and range fields.

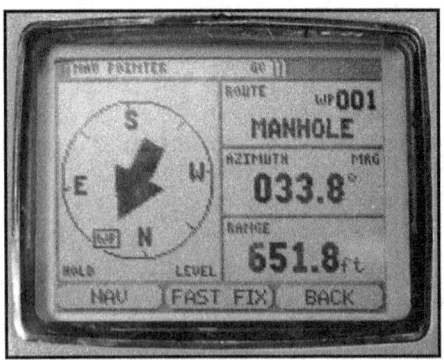

Figure K-6. NAV pointer.

(4) ***Map***. The Map page displays a graphical map of relationships between current position, landmarks, map objects, and selected waypoints. With a map previously loaded, the Map page automatically displays a map with present position of the DAGR shown at the center of the display. The operator uses zoom and pan operations, and waypoint selections to obtain a desired view. When navigating, the Map page provides the operator with a mapped view of surrounding terrain and potential obstructions (for example, a body of water).

(5) ***SV Sky View*** (Figure K-7). The SV Sky View page displays status information on tracked satellites (for example, acquiring satellites). The current operating status is shown at the top of the display. Numbers inside black circles indicate satellites in use to acquire or maintain the current DAGR position. The corresponding number at the left side of the display provides a bar graph indication of satellite signal strength and code status. The longer the bar, the greater the signal strength. A black bar indicates ephemeris data is collected. If the DAGR cannot display satellite information, no bars appear at the far left side of the display.

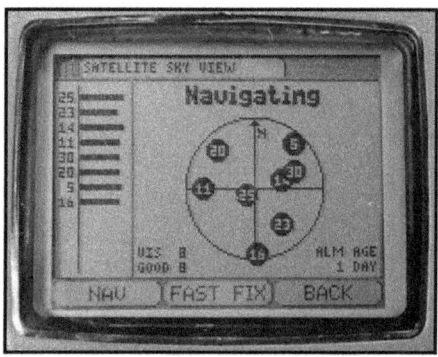

Figure K-7. SV sky view.

K-3. OPERATIONAL CHECKOUT PROCEDURES

Performing the operational checkout procedures on the DAGR determines if the unit is operating correctly. These checks aid the user in detecting a DAGR malfunction that may be corrected in the field. If the DAGR passes the operational checkout procedure, the unit is ready to use; if the DAGR does not pass the operational checkout procedure, proceed to the troubleshooting procedures. After operational checkout procedures, but before use, ensure:

- The correct function set is being used (basic or advanced).
- The correct user profile is being used if using the advanced function set.
- The DAGR is set to the desired operating mode.

a. **External Body**. Inspect the DAGR and external cables and equipment for damaged and or missing parts.

b. **Power Up**. Push the POWER key to turn the DAGR on, and make sure the DAGR has a clear view of the sky. During power up, observe the power-on status display. Make sure the DAGR passes the self-tests and the battery strength indicator shows sufficient battery power remaining. Do not use the DAGR if a failed self-test is indicated.

NOTE: A Test Summary page can be accessed from the System submenu for a listing of tests that passed or failed. Do not use the DAGR if the Status field shows FAILED.

c. **Operator-Induced Commanded Self-Test.** After the DAGR successfully completes the power-on self-test and shows the SV Sky View page or Present Position page, perform the following procedure for an operator-induced commanded self-test.

NOTE: The self-test does not track SVs, determine position, or provide navigation data. Operator confirmation is required to enter this mode. The self-test lasts approximately four minutes and requires operator intervention to complete.

(1) Activate commanded DAGR self-test.

(2) If the Present Position page is not already displayed, push and hold the POS key (except when showing a message pop-up, then push the QUIT key first). The Present Position page is displayed. Push the MENU key.

(3) Highlight Select Op Mode, then push the ENTER key.

(4) Highlight Test, then push the ENTER key.

(5) The DAGR displays an ENTER TEST MODE message prompting the operator to confirm or cancel entering test mode. Push ENTER key to confirm.

(6) A Test In Progress display appears with the specific area of testing listed at the bottom and a bar graph denoting progress. The DAGR automatically tests multiple areas.

NOTE: While performing the following keypad test, push and hold the ENTER key to test the ENTER key. Push and release the ENTER key to advance to the next display.

(7) After the displayed tests are completed, Keypad Test is displayed. Push each key on the keypad and verify the corresponding key shown on the display toggles between normal and highlighted appearance. Push the ENTER key to continue to the next display.

(8) The Display Light Test display appears with the brightness adjustment cycling between 0% and 100%. The percentage adjustment is reflected in the light bulb of the display. Push the ENTER key to continue.

(9) The Contrast Test display appears with the contrast adjustment cycling between 0% and 100%. The percentage adjustment is reflected in the bar graph of the display. Push the ENTER key to continue.

(10) The Display Test Beginning message appears momentarily. After sequencing through white, light gray, dark gray, and black, the Display Test Completed message appears, followed by the Power-On Status display listing self-test results as Pass or Fail.

NOTE: A Test Summary page can be accessed from the System submenu for a listing of tests that passed or failed. Do not use the DAGR if the Status field shows FAILED.

(11) If the Power-On Status remains displayed and does not time out, push the ENTER key to acknowledge.
(12) The SV (satellite vehicle) Sky View page is displayed. Push the MENU key.
(13) Highlight Select Op Mode, then push the ENTER key.
(14) Highlight Continuous, then push the ENTER key. This mode enables the DAGR to acquire a current position fix.
(15) After satellites are acquired and a current position fix is obtained, the DAGR display stops blinking and Tracking SVs is shown on the SV Sky View page. The display then automatically switches to the Present Position page.

NOTE: If the DAGR does not acquire satellites, the display blinks between black and gray text and goes into Standby mode (both handheld and host platform operation). If the DAGR display continues to blink, verify a clear view of the sky, then perform the manual initialization procedure.

K-4. BATTERY INSTALLATION

The DAGR requires two sets of batteries to operate fully, although it can operate on only the primary battery. Table K-3 shows the types of batteries the DAGR can use.

a. **Battery Life and Type**. The approximate battery life is based on operating the DAGR in continuous mode, at room temperature, and without keypad/display lighting. Several operator selectable DAGR settings are available to extend battery life. No power conservation is required when using external power. Internal batteries are not required when using external power, and need not be removed when connected to external power

Type	Size and Voltage	Use	Life	Rechargeable ?
Lithium	AA 1.5 volt	Primary	16.5 hours	No
Alkaline	AA 1.5 volt	Primary	11.5 hours	No
Alkaline	AA 1.5 volt	Primary	7 hours	Yes
Nickel Metal Hydride	AA 1.5 volt	Primary	10 hours	Yes
Lithium	1/2 AA 3.6 volt	Memory	6 months	No

Table K-3. Types of batteries.

WARNING

If abused, lithium batteries can explode causing severe injury. Be sure to store batteries in original packaging until ready to use, and observe polarity during installation. Reverse polarity can cause damage to the battery and receiver.

> **CAUTIONS**
> 1. Do not mix new batteries with old batteries. Do not mix battery types. Do not reverse battery polarity. Use only fresh/new batteries.
> 2. If the DAGR is being used for the first time and there are no memory settings to be saved, the memory battery is not important, but still needs to be installed. To ensure all settings from previous usage are retained, ensure a good memory battery is installed (check memory battery date on Battery page) or external power is applied to the unit before installing or replacing the primary batteries.
> 3. If all primary and memory power is lost, memory information is lost, and the DAGR resets to default settings after power-up.

   b. **Primary Battery Installation.** Use the following steps to install the primary battery.
   (1) Ensure power to the DAGR is off.
   (2) Hold the unit firmly upside down with the battery pack facing up.
   (3) Push or pull the latch located on the battery pack to release the battery pack. Lift up on the battery pack and remove it from the unit.
   (4) Position the battery removal strap into the channel of the battery pack before installing new batteries.
   (5) Install new batteries and ensure correct polarity installation for each battery (marked on battery pack).
   (6) Before installing the battery pack, inspect the battery pack gasket for damage or dirt. Lubricate or replace gasket if necessary. Ensure the battery removal strap is not protruding from the battery pack.
   c. **Memory Battery Installation.** To install a new battery pack, position the tab on the battery pack in the slot on the DAGR.
   (1) Close the battery pack against the DAGR until the battery pack is engaged.
   (2) Ensure power to the DAGR is off.
   (3) Place the unit upside down on a nonabrasive surface with the memory battery cover facing up.
   (4) Use a flat-blade screwdriver to loosen the three screws securing the memory battery cover, then remove the cover from the unit.
   (5) Remove the expired memory battery and properly dispose of it.
   (6) Install the new memory battery.
   (7) Before reinstalling the memory battery cover, inspect the gasket for damage or dirt. Lubricate or replace the gasket if necessary.
   (8) Reinstall the memory battery cover, and tighten the three screws.

K-5. OPERATING PROCEDURES

Operating procedures include the following steps:
- Turn the power on.
- Conduct a self-test.
- Select the mode of operation.
- Turn the power off at the end of the operation.

a. **Preoperational Steps and Checks.** Several steps and checks must be conducted before turning the DAGR on. (Refer to TM 11-5820-1172-13 for more details.)

(1) To ensure proper battery life and proper unit operation, check the batteries to make sure they are of the same type, are not a mix of old and new batteries, and are still good (by checking the battery indicator).

(2) If using external power, be sure the battery cable is properly connected.

(3) Be sure the DAGR has an open view of the sky to acquire the present position. When position data fields blink between black and gray text, the DAGR is not tracking satellites or has not yet acquired present position.

(4) Manually enable and orient the internal compass.

(5) To operate in -20-degree Centigrade or below conditions, the heater must be on for at least 20 minutes before powering up.

(6) If a warning or other message displays while operating the DAGR, follow the display instructions.

b. **Turn the Power On.** Push the PWR key to turn the DAGR on. A display page briefly appears indicating the DAGR software version. After the power is on, the normal operating mode is Continuous when operating on external power and Fix when operating on battery power.

(1) If a CV key, group-unique variable (GUV) key, or an SV code condition exists, acknowledge the message(s) accordingly.

(2) The power-on status message (Figure K-8, page K-14) appears and provides the following information. (All messages may not be listed as they are dependent on how the DAGR is configured. Use the up/down cursor control keys to scroll and view the entire display message.)

(a) *Self-Test.* The Self-Test message displays the self-test results as PASS (no self-test failures found) or FAIL (self-test failures detected). This message is always displayed.

(b) *Battery Used.* The Battery Used message indicates the primary battery capacity used (amount of time in hours and minutes the DAGR has been operated using primary battery). It is displayed when using internal primary battery power only.

(c) *Battery Left.* The Battery Left message indicates the primary battery capacity remaining (in hours and minutes). It is displayed when using internal primary battery power only.

(d) *CV.* The CV message indicates whether or not CV keys are loaded. If they are loaded, the message indicates whether or not the DAGR has the current CV key.

(e) *Power.* The Power message indicates which power is being used—internal or external.

(f) *Days Remaining.* This message indicates the number of days remaining in the mission. It also indicates if enough CV keys are loaded for mission duration.

(g) *Default.* The Default message indicates that the DAGR's position, time, and date are default values. It also indicates if initialization is recommended for the DAGR.

Figure K-8. Power-on status.

(3) The Power-On Status message times out in two seconds. The DAGR is ready for use if the status message indicates that the self-test has passed and the DAGR does not need initialization. If the self-test indicates FAIL, the operator is prompted to push the ENTER key to acknowledge, but the DAGR is not ready to use. If any of the following conditions exist, a message requiring the operator's acknowledgement will appear.

- No CVs or GUV keys are loaded.
- No CV key for today is loaded.
- Not enough CV keys are loaded for duration of mission (if mission duration is entered).
- SV code is set to Mixed.

(4) After the Power-On Status display times out or is acknowledged, the DAGR displays the SV Sky View page with satellite acquisition status appearing at the top. Initially, the status is displayed as ACQUIRING SVS and then, when successful, the status changes to TRACKING SVS.

(5) After the DAGR has acquired the current position, the unit automatically switches to the Present Position page and displays the position coordinates, elevation, and the estimated horizontal error (EHE).

(6) During satellite acquisition, the PAGE or QUIT keys can be used to access the Present Position page. If the DAGR is not tracking satellites, the display will blink and the Present Position page will display the last position recorded by the receiver before being turned off.

NOTE: Adjust the DAGR Keypad/Display lighting is by pushing and holding the BRIGHTNESS key and the respective up or down cursor control key.

c. **Conduct a Self-Test.** Conduct the self-test using the same procedures described in paragraph K-3.

d. **Mode of Operations.** The DAGR mode of operation can be selected from any display, except a message pop-up, by pushing the MENU key twice to display the Main Menu. The normal (default) operating mode is Continuous when the DAGR uses external power and Fix when it uses battery power.

(1) *Types of Modes.* The DAGR can operate in eight different modes.

(a) *Continuous.* The Continuous mode, which uses the most power, tracks satellites to produce a continuous PVT solution.

(b) *Fix.* The Fix mode tracks satellites to produce a current PVT solution. After a position fix is obtained, it automatically transitions to the Standby mode.

(c) *Standby.* When the DAGR is in the Standby mode, which uses reduced power, it does not acquire and track satellites, but performs all functions that do not require satellites.

(d) *Other Available Modes.* The other available modes are: Average, Time Only, Rehearsal, Test, and Off. (Refer to TM 11-5820-1172-13 for details on their use.)

(2) *Selection of the Operating Mode* (Figure K-9). To select an operating mode—

(a) From any display except a message pop-up, push and hold the POS/PAGE key until the Present Position page is displayed.

(b) Push the MENU key.

(c) Highlight the Select Op Mode option and then push the ENTER key.

(d) Highlight the desired operating mode and then push the ENTER key.

(e) The display returns to the Present Position page and displays the current operating mode.

Figure K-9. Operating modes.

e. **Power Off.** The user performs the following operations to turn the DAGR off after use.

(1) Push and hold the PWR/QUIT key. The power down warning page is displayed. If some functions are enabled, such as Auto-on, a message appears before the power down warning.

(2) Wait the allotted time for the DAGR to turn off, or push the WP/ENTER key to turn it off immediately.

K-6. MANUAL INITIALIZATION

If the DAGR has been moved a long distance and is not operating properly, it may need to be initialized according to its new position. Some indications that it needs to manually initialized include:

- Difficulty obtaining a position fix.
- Datum is mismatched with navigation waypoints.
- Datum does not match the geographical map being used.

(Refer to TM 11-5820-1172-13 for instructions on performing a manual initialization.)

K-7. CREATION OF WAYPOINTS

A waypoint is a position reference used to navigate, define routes, or mark points of interest. The Waypoints page is accessed using the WP key or from the WP/Routes/Alerts submenu. The Waypoints page provides a table that lists all DAGR waypoints. The Waypoints Editor page allows the user to create new waypoints, edit existing waypoints, clear waypoints, copy waypoints, or view only desired waypoints (search, sort, and filter). The Waypoint Editor page is accessed from the Waypoints page.

a. The menu functions on the Waypoints page are (Figure K-10):

(1) *Create/New.* This function provides a list of unused waypoints (numbers). After selecting a new waypoint number, the Waypoint Editor page is used to set up the waypoint.

(2) *Edit Waypoint.* This function displays the Waypoint Editor page for editing the selected waypoint.

(3) *Copy.* The copy function copies a selected waypoint's data. This data can then be pasted into another waypoint or a range of waypoints. Operator confirmation is required before the DAGR overwrites any existing waypoints.

(4) *Clear.* The clear function clears a single waypoint, a range of waypoints, or all waypoints. Operator confirmation is required before the DAGR clears any waypoints.

(5) *Units.* This function provides an editor to select range, angle, north reference, or elevation (Advanced) units.

(6) *Navigate to Waypoint.* This function displays the NAV Pointer page.

(7) *Search.* The search function searches for and displays waypoints by a name or remark (up to ten characters each).

(8) *Sort.* The sort function sorts and displays the entire list of waypoints in ascending alphanumeric order by name, number, range from present position, range from selected waypoint, or identity.

(9) *Filter.* The filter function displays a filtered list of waypoints. Filter choices are: All Used WPs, All Unused WPs, Within Range (specified by operator), and Unfilter (display all waypoints).

(10) *Waypoint Summary.* The waypoint summary displays the number of waypoints used and unused.

FM 3-25.26

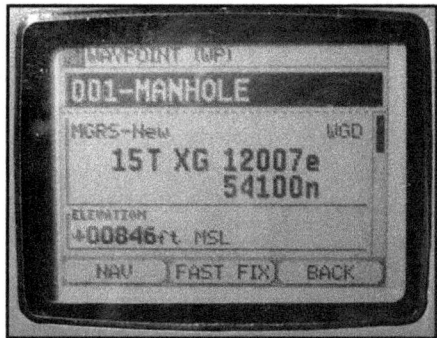

Figure K-10. Creating waypoints.

b. Perform the following steps to create a waypoint.

(1) From any display, push and hold the WP key. Waypoint function choices are displayed.

(2) Highlight Create New WP, then push the ENTER key.

(3) The Waypoint Editor page automatically displays the first unused waypoint with current (if tracking satellites) or last position information. Revise information as necessary.

(a) Use the left and right cursor keys to move between the digits.

(b) Use the up and down keys to change the digit, and push ENTER to enter the new digit.

c. Complete the action by highlighting one of the following and pushing the ENTER key.

(1) *Save and Exit.* The display briefly shows the Waypoint Stored message, then returns to the Waypoints page with the new waypoint information saved and highlighted.

(2) *Exit and No Save.* When this option is selected, the display returns to the Waypoints page without saving the waypoint.

(3) *Edit Field.* This selection displays an editor for the highlighted field.

(4) *Undo Changes.* Undo Changes clears any changes made, and the display returns to the Waypoints Editor page for editing.

(9) *Help.* This selection displays help text for the highlighted field.

K-8. CREATION OF A NEW ROUTE

By creating a route, the operator creates or designates existing waypoints to define the course he wants to take. Waypoints define the end of each leg of the route and the sum of the designated legs becomes the route. The Route Editor page allows the operator to view, edit, and create routes. The Routes page is accessed from the WP/Routes/Alerts submenu. It provides a table that lists all routes stored in DAGR. Vertical scrolling is used to view all routes. If a route is undefined or invalid, double dashes appear in the route name and legs columns of the table. The route list includes the following information for each route:

- **NUM** — Displays the route number (01 through 15).
- **Route Name** — Displays the route name of up to ten characters.

- **Legs** — Displays the number of route legs (1 to 1,000). This quantity matches the number of waypoints in a route.

a. From the Routes page, highlight the desired new route number, or if the highlighted route is not changed, the first unused route number will be automatically used. Push the MENU key. Highlight Create and push the ENTER key.

b. Highlight Create/New and push the ENTER key. The Route Editor page displays the first unused route if no route number was previously selected or the operator-selected route.

c. Scroll down the page into the route leg table and highlight the first row containing all double dashes (unused leg), then push the ENTER key. Highlight the desired ending waypoint for the leg of the route, then push the ENTER key. The Route Editor displays the route leg with the selected end waypoint. Repeat this step to create all desired route legs. After creating all route legs, push the MENU key.

NOTE: The top row of the route leg table always has end waypoint 000–POS representing present position.

d. Complete the action by highlighting one of the following and pushing the ENTER key.

(1) *Save and Exit.* The display briefly shows the route stored message, then returns to the Routes page with the new route information saved and highlighted.

(2) *Exit and No Save.* The display returns to the Routes page without saving the route.

(3) *Maximize/Minimize Table.* The display returns to the Routes page with the route table maximized (displaying five routes at once) or minimized (displaying three routes at once).

(4) *Insert WP After.* From the Select WP editor, highlight the desired waypoint, then push the ENTER key. The Route Editor page highlights a new leg (inserted after the original highlighted leg) created from the entered waypoint.

(5) *Swap With Next.* The Route Editor displays the highlighted route leg swapped with the one that was next (disabled if the highlighted route leg is the last leg).

(6) *Remove WP.* The Route Editor page displays with the highlighted leg removed.

(7) *Edit Field.* This selection displays an editor for the highlighted field (leg) (Figure K-11).

(8) *Undo Changes.* Undo Changes clears any changes made, and the display returns to the Route Editor page for editing.

(9) *Help.* This selection displays help text for the highlighted field.

Figure K-11. Create a route.

e. To calculate the length of a route, highlight a desired route from the Routes page, and push the MENU key.
(1) Highlight Calculate Length, then push the ENTER key.
(2) The route length is displayed. Push the ENTER key to acknowledge.
(3) The DAGR returns to the Routes page.

K-9. CREATION OF A NEW ALERT
Setting the DAGR's alert function will indicate to the operator when he is approaching or has passed a waypoint (Figure K-12) or a predefined line or area. The Alerts page is accessed from the WP/Routes/Alerts submenu, or by using the Status key and Receiver Status menu. The Alerts page provides a table showing all DAGR alerts. The operator can create new alerts, edit existing alerts (using Alert Editor page), clear alerts, copy alerts, and enable or disable alerts. The Alert Editor page is accessed from the Alerts page. Vertical and horizontal scrolling is used to view all alerts and table columns. If alert data are undefined or invalid, double dashes appear in the table columns. Use the Status key to view the Receiver Status display and check the alert status or access the Alerts page.

Figure K-12. Alert message.

a. From the Alerts page, highlight the desired new alert number, or if the highlighted alert is not changed, the first unused alert number will automatically be used. Push the MENU key.

b. Highlight Create/New, then push the ENTER key. The Alert Editor page displays the first unused alert (up to 33 alerts can be created) if no alert was previously selected. Revise the information as necessary using standard editing techniques, then push the MENU key. Highlight the desired option from the multiple options provided, then push the ENTER key.

d. Complete the action by highlighting one of the following and pushing the ENTER key:

(1) *Save and Exit.* The display briefly shows the alert stored message, then returns to the Alerts page with the new alert information saved and highlighted.

(2) *Exit and No Save.* The display returns to the Alerts page without saving the alert.

(3) *Edit Field.* This selection displays an editor for the highlighted field.

(4) *Undo Changes.* Undo Changes clears any changes made, and the display returns to the Alerts Editor page for editing.

(5) *Help.* This selection displays help text for the highlighted field.

K-10. DAGR NAVIGATION

The operator usually navigates directly to a waypoint or follows a route with legs created by moving to a series of waypoints. He can also use the Elevation Hold and Bullseye method.

a. **Selection of the Navigation Function.** Push the MENU key twice, highlight Navigation, and push the ENTER key. Scroll to Navigation Setup and push the ENTER key. Scroll to the Navigation Method field and push ENTER. Select either Direct To or Route. (Although other methods are available, such as Course To, Approach, and so on, Direct To and Route are the most common and are discussed herein. Refer to TM 11-5820-1172-13 for the other methods.)

(1) *Direct To.* Set the To WP field to the waypoint being navigated to.

(a) Highlight To WP field, then push the ENTER key.

(b) Highlight the desired waypoint and push the ENTER key.

(c) Set the WP Alert Mode field to On or Off. When the Alert Mode field is on, the DAGR visually alerts the operator upon arrival at the waypoint.

(d) The Calc Type field appears on all NAV Setup page displays when using advanced function set. There are two methods used for calculating navigation information: Rhumb Line (RL) or Great Circle (GC).

- **Rhumb Line (RL)**—produces constant compass directions and allows lines of latitude to be used as paths.
- **Great Circle (GC)**—produces the shortest path to the navigation waypoint, but the compass direction of travel changes due to the curvature of the earth.

(4) *Route.* Perform the following steps to navigate using a series of waypoints along a route that has been previously created.

(a) Highlight Route and push the ENTER key.

(b) Set the Route field to the desired navigation route number/name.
- Highlight Route field, then push the ENTER key.
- Highlight the desired route and push the ENTER key.

(c) Configure the Calc Type Field and the alert function the same as for the Direct To method.

b. **Navigating with the DAGR.** Access the NAV Pointer page, then travel the azimuth pointed by the Pointer field arrow. The compass dial rotates so the top of the dial indicates the current ground track.

(1) If the DAGR internal compass is active, Hold Level appears at the bottom of the Pointer field. The internal compass activates when moving below a preset speed for a preset amount of time.

(2) While moving towards the destination waypoint, the Range field value steadily decreases and the Azimuth field value changes.

(3) Use the Steering 2D field directional arrow and angular value to align the track with the azimuth for navigation to the leg ending waypoint. When off course, the left and right directional arrows and angular value appear. When on course, the on course indicator (.....) appears, and the Track field and Azimuth field values match.

(4) At any time during route navigation, the operator can reverse the direction of travel on the route by changing the setting of the Direction field of the NAV Setup page.

(5) The DAGR recognizes it has reached the waypoint when it reaches a radius from the waypoint (default is 5 meters) set in the Alert Radius field of the Waypoint Editor page. The operator must confirm waypoint arrival only if the WP Alert Mode field of the NAV Setup page was previously set to On. If the operator is using the Route function, the DAGR automatically switches to the next leg.

(6) Set the Direction field to Forward or Reverse for desired direction of navigation through the route legs.

K-11. SITUATIONAL AWARENESS

The Situational Awareness page provides a graphical display of the DAGR's current position compared to other waypoints, routes, and alerts shown on the display. It is accessed from the NAV submenu (Figure K-13, page K-22). Track, ground speed, north indicator, position error, and range scale data all provide additional DAGR present position information. The operator can select view orientation, view content (waypoints, routes, and alerts), edit displayed waypoints, measure between selected points, and track history.

FM 3-25.26

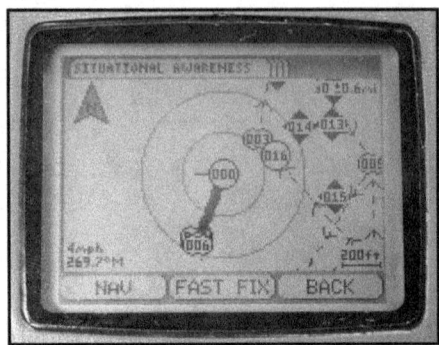

Figure K-13. Situational awareness.

a. The present position symbol (waypoint 000 inside a circle) is at the center of the display (unless fields are selected) with a track indicator staff.

(1) Ground speed and track are displayed in the lower left corner.

(2) If the DAGR internal compass is being used, displayed track text alternates with the instruction to HOLD LEVEL.

(3) The range scale is displayed in the lower right corner. The north reference indicator is displayed in the upper left corner and always points to True North.

(4) Position error (EHE, EPE, EVE, or FOM) is displayed as a ± value in the upper right corner (except FOM is displayed as a value of 1 to 9, with 1 being the best).

(5) The operator can set the display view orientation as follows:

- North-Up—top of the display is north.
- Track-Up—top of the display is current track.
- Course-Up—top of the display is current navigation course (if defined, otherwise defaults to current track).
- Operator-Entered—top of display is an operator-entered **value.**

b. The Situational Awareness page provides a graphical display of waypoints, routes, alerts, and track history. Page characteristics include zoom range scale, panning/scrolling, and measuring range between points.

(1) *Waypoints.* Waypoint numbers are provided in the center of a shape and a direction indicator staff is attached, if applicable. Highlighted symbols (shown on the bottom row) denote selected waypoints. The operator can select which waypoints to display.

(2) *Routes.* Routes are shown as dashed lines with arrows indicating route direction from waypoint to waypoint (legs of the route). The display of waypoints used to define the route is based upon the selected waypoint view option. The operator can select which routes are displayed.

(3) *Alerts.* Waypoints are used to define alerts. Alerts are displayed as selected by the operator (none, enabled, or all) using shapes to denote the alert type. Spikes displayed as part of the alert perimeter represent the dangerous side or area of an alert. Use the Status key and the Receiver Status menu to check alerts and their status.

(4) *Zoom/Range Scale.* The operator can zoom in or out using the IN or OUT keys on a scale of 50 feet to 800 miles, 50 yards to 800 nautical miles, or 50 meters to 800

kilometers (English, nautical, or metric units). Range scale is shown in the lower right hand corner of the display.

(5) *Overzoom.* Overzoom is displayed in place of the range scale when the DAGR speed is too fast for the selected zoom scale. The operator may zoom out until overzoom is no longer displayed. When zooming in or out with the cursor displayed, the display centers upon the cursor. When a waypoint is selected, the display centers upon the waypoint and the cursor moves to the center of the display.

(6) *Panning and Scrolling.* The operator uses the cursor control keys to pan (move) the display to any horizontal point. Default panning is the present position (POS) at the center of the display. A cursor appears when scrolling in any direction. Scrolling the cursor to the edge of the view pans the display.

(7) *Measuring Range Between Points.* When the measurement function of the page menu is used, a measurement box appears. The measurement box provides azimuth (AZ), range (RNG), slant range (SR), and elevation angle (ELA) data computed from a starting point position (DAGR position) to the current cursor position. A corresponding line between the points also appears. The operator can set the starting point as a point other than present position and restart the measurement. When the cursor is moved to a waypoint, the waypoint symbol is highlighted to signify the waypoint is selected for generating measurement data.

c. The Map page is accessed from the Navigation submenu (Figure K-14). The Map page provides a graphical map display of relationships between current position, landmarks, map objects, and selected waypoints. With a map previously loaded (covering present position), the Map page automatically displays a map with the DAGR's present position shown at the center of the display. The operator uses zoom and pan operations, and waypoint selections to obtain a desired view. When navigating, the Map page provides the operator with a mapped view of surrounding terrain and potential obstructions (for example, a body of water).

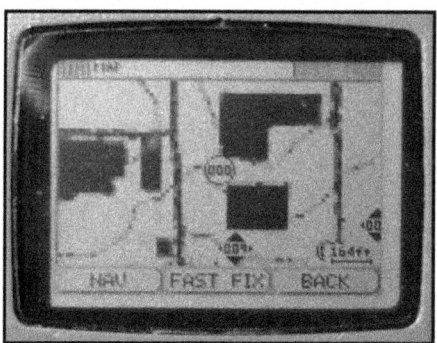

Figure K-14. Map page.

(1) The Map page is always oriented with the top of the map appearing at the top of the display.

(2) The present position symbol (waypoint 000 inside a circle) is at the center of the display (unless the operator is panning the display).

(3) A scale value (dependent upon the map in use) is provided in the lower right corner of the Map page.

(4) The Map page uses no fields and no cursor.

(5) The Map page provides a graphical display of waypoints and map objects. Page characteristics include zoom range scale, panning, and map selection.

(6) Waypoint numbers are displayed in the center of a shape denoting their identity or type as follows:
- Friendly—circle.
- Hostile—diamond.
- Neutral—square.
- Unknown—cloud.

Present position is always displayed as waypoint 000.

(7) The operator can select which waypoints to display from the Map page menu. Operator selectable waypoint view options are:
- None—only present position is displayed.
- Navigation—displays waypoints being used for navigation inside the map coverage area as determined by the NAV Setup page From WP or To WP fields.
- Operator-selected—displays waypoints inside the map coverage area that have been selected by the operator for viewing.

K-12. TROUBLESHOOTING

Troubleshooting procedures detect and isolate DAGR failures and malfunctions. These procedures are similar to the operational checkout procedures. After a DAGR failure has been found and corrected, perform the operational checkout procedure again to make sure the DAGR is operating properly. If troubleshooting confirms a DAGR failure, and repair is beyond what is covered in the DAGR TM, turn the DAGR unit in.

a. **Inspect External Body.** Inspect the DAGR and external cables and equipment for damage and or missing parts. If the DAGR is damaged or parts are missing, turn the unit in.

b. **Power Up the DAGR.** Push the PWR key to turn the DAGR on.

c. **Check Battery.** If the DAGR display does not come on after power is applied, check the primary battery and primary battery pack, and the external connections to the power source. If the battery connections are good, turn the unit in.

(1) If operating in cold conditions, allow additional time (up to 20 minutes) for the display to appear.

(2) Observe the power-up test results. If a failure occurs, check all external connections and rerun the self-test. Follow instructions on the display or turn the unit in.

(3) After power-up when using primary power batteries, check the primary battery life indicator to determine if the battery has sufficient life. If battery life is insufficient, replace primary batteries and update battery information on the Battery page.

(4) After power-up, ensure a low memory battery message does not show. Access the Battery page and check the date shown in the Memory Battery Installed field. If the DAGR shows a date approximately six months old, then replace the memory battery and update the memory battery information on the Battery page.

d. **Perform an Operator-Induced Commanded Self-Test.** After power-up, perform an operator-induced commanded self-test.

(1) If the DAGR fails the self-test, check all external connections if applicable, and rerun the self-test.

(2) Follow instructions on the display and or turn the unit in.

e. **Display Blinks Between Black and Grey Text.** After power-up, if the DAGR passes the self-test and the display still blinks between black and grey text, perform the following:

- Move the DAGR (or external antenna) to an open view of the sky.
- Hold the DAGR at a 90-degree angle to the horizon.
- Ensure satellite acquisition time was at least two minutes.
- Ensure the DAGR is in a satellite tracking mode of operation.
- Perform the manual initialization procedure.

(1) If the display does not stop blinking, turn the unit in.

(2) If the display stops blinking, troubleshooting is complete.

This Page intentionally left blank.

GLOSSARY

AA	avenue of approach
ANCOC	Advanced Noncommissioned Officer Course
AR	Army regulation
BM	bench marks
BNCOC	Basic Noncomissioned Officer Course
BT	basic training
*CCC	Captain's Career Course
cm	centimeter
CONUS	continental United States
CS	combat support
CSS	combat service support
CUCV	commercial utility cargo vehicle
DD Form	Department of Defense form
E	east
EPLRS	Enhanced Position Location Reporting System
FIST	fire support team
FM	field manual
FORSCOM	United States Army Forces Command
GD	ground distance
GEOREF	geographic reference
G-M	grid-magnetic
GPS	Global Positioning System
GSR	ground surveillance radar
GTA	graphic training aid
G/VLLD	ground/vehicular laser locator designator
HD	horizontal distance
HHC	headquarters and headquarters company
HMMWV	high-mobility multipurpose wheeled vehicle
JOG	joint operations graphics
JTIDS	Joint Tactical Information Distribution System
km	kilometer
LAT	latitude

MD	map distance
METT-TC	mission, enemy, terrain and weather, troops and support availab time available, civil considerations
MITAC	Map Interpretation and Terrain Association Course
N	north
NCO	noncommissioned officer
NGA	National Geospatial-Intelligence Agency
OAC	Officer Advanced Course
OBC	Officer Basic Course
OCS	Officer Candidate School
OSUT	one station unit training
PADS	Position and Azimuth Determining System
PD	photo distance
PJH	hybrid (PLRS and JTIDS)
PLGR	precision lightweight Global Positioning System receiver
POI	program of instruction
PRE	precommission
QRMP	quick response multicolor printer
RF	representative fraction
ROTC	Reserve Officers' Training Corps
S	south
SF	standard form
SME	subject matter expert
SOSES	shape, orientation, size, elevation, and slope
SUSV	small-unit support vehicle
tan	tangent
TM	technical manual
TOW	tube-launched, optically tracked, wire-guided missile
TRADOC	United States Army Training and Doctrine Command
topo	topographic
UPS	universal polar stereographic
U.S.	United States
USGS	United States Geological Survey
UTM	universal transverse mercator
VD	vertical distance
VNAS	Vehicular Navigation Aids System
W	west
*WLC	Warrior Leader Course

REFERENCES

SOURCES USED
These are the sources quoted or paraphrased in this publication.

FM 7-0	Training the Force. 22 October 2002.
FM 7-1	Battle Focused Training. 15 September 2003.
FM 21-31	Topographic Symbols. 19 June 1961.
FM 25-4	How to Conduct Training Exercises. 10 September 1984.
FM 101-5-1	Operational Terms and Graphics. 30 September 1997.
TM 11-5820-1172-13	Operator and Maintenance Manual for Defense Advanced GPS Receiver (DAGR) Precise Positioning Service (PPS). 06 June 2004.
TM 11-5855-238-10	Operator's Manual for Night Vision Goggles Ground Use: AN/PVS-5 and ANPVS- 5A. 15 May 1993.
TM 11-5855-262-10-1	Operator's Manual for Night Vision Goggles Ground Use: ANPVS-7B and ANPVS-7D. 01 July 1994.

DOCUMENTS NEEDED
These documents must be available to the intended users of this publication.

AR 115-11	Geospatial Information and Services. 10 December 2001.
AR 380-5	Department of the Army Information Security Program. 29 September 2000.
AR 380-40	Policy for Safeguarding and Controlling Communications and Security (COMSEC) Material. 30 June 2000.
DD Form 1348	DOD Single Line Item Requisition System Document (Manual). July 1991.
DD Form 1348M	DOD Single Line Item Requisition System Document (Mechanical). March 1974.
FM 5-33	Terrain Analysis. 11 July 1990.
FM 6-40	Tactics, Techniques, and Procedures for Field Artillery Manual Cannon. 24 April 1996.

FM 3-25.26

FM 34-1 Intelligence and Electronic Warfare Operations. 27 September 1994.

FM 34-3 Intelligence Analysis. 15 March 1990.

GTA 05-02-012 Coordinate Scale and Protractor. 02 January 1981.

TM 11-5825-291-13 Operations and Maintenance Manual for Satellite Signals Navigation Sets AN/PSN-11 and AN/PSN-11(V). 01 April 2001.

TB 11-5825-291-10-2 Soldiers' Guide for the Precision Lightweight GPS Receiver (PLGR) AN/PSN-11. 1 December 1996.

TB 11-5825-291-10-3 The PLGR Made Simple. 1 November 1997.

INTERNET WEBSITES

Reimer Digital Library, http://www.adtdl.army.mil

Army Publishing Directorate, http://www.usapa.army.mil

INDEX

aerial photographs
  comparison with maps, 8-1
  features, identification of, 8-18
  film, 8-7
  indexing
    four-corner method, 8-11 (illus), 8-12 (illus)
    template method, 8-12 (illus), 8-13 (illus)
  map substitute, as, 2-7
  marginal information, recording, 7-3, 7-4 (illus)
  numbering, 8-7
  orienting, 7-3, 8-13, 8-14 (illus)
  point designation grid, 8-15 (illus) to 8-18 (illus)
  scale determination, 8-8
    comparison method, 8-9 (illus)
    focal length-flight altitude method, 8-10 (illus)
  stereovision, 8-19 to 8-23 (illus)
    mirror stereoscope, 8-22 (illus)
    pocket stereoscope, 8-21, 8-22 (illus)
  titling, 8-7
  types
    convergent, 8-7
    high oblique, 8-5 (illus), 8-6 (illus)
    low oblique, 8-3, 8-4 (illus)
    multiple lens, 8-7
    panoramic, 8-7
    trimetrogon, 8-6 (illus)
    vertical, 8-1 to 8-3 (illus)
AN/PSN-11 (*see* PLGR)
AN/PVS-5/5A, night vision goggles, G-1 (*see also* navigation aids)
AN/PVS-7B/D, night vision goggles, G-1 (*see also* navigation aids)
arctic terrain (*see also* terrain, navigation in different types of)
  interpretation and analysis, 13-9
  navigation, 13-10
  operations, 13-9

azimuths, 6-2 (illus)
  back, 6-3 (illus)
  compass, preset, and follow an azimuth, 9-4 to 9-6 (illus)
  grid, 6-3, 6-4 (illus)
  magnetic, 6-3

bar scales (*see* distance: graphic [bar] scales)
base lines
  grid north, 6-2 (illus)
  magnetic north, 6-2 (illus)
  true north, 6-1, 6-2 (illus)
bench marks, 10-5

catching feature, 11-18
cliff, 10-14, 10-15 (illus) (*see also* terrain features)
compasses
  handling, 9-2
    accuracy, 9-3
    inspecting, 9-2
    metal and electricity, effects of, 9-2
    protection, 9-3
  lensatic, 9-1 (illus), 9-2
  M2 (*see* M2 compass)
  types, 9-1
  use
    bypass an obstacle, 9-6 (illus)
    centerhold technique, 9-3, 9-4 (illus)
    compass-to-cheek technique, 9-4 (illus)
    offset, 9-6, 9-7 (illus)
    orienting, 11-1 to 11-3 (illus), 11-4 (illus)
    preset and follow an azimuth, 9-4 to 9-6 (illus)
contour intervals, 10-2 (illus) to 10-5 (illus)
contour lines, 10-1
  index, 10-2 (illus)
  intermediate, 10-2 (illus)

supplementary, 10-2 (illus)
terrain, interpreting, 10-17
conversion tables, C-1 (table) to C-3 (table)
coordinates
   geographic, 4-1 to 4-9 (illus), 4-10 (table)
   grid, 4-14 to 4-17 (illus)
   map, protection of, 4-26
cut, 10-15, 10-16 (illus), E-19 (illus) (*see also* terrain features)

DAGR (defense advanced GPS receiver) (*see also* navigation aids)
   battery installation
      life and type, K-11, K-12 (table)
      memory, K-13
      primary, K-12
   checkout procedures, operational
      external body, K-10
      operator-induced commanded self-test, K-10
      power up, K-10
   controls and indicators
      display windows, K-3 to K-5 (illus)
      editors, K-5 to K-7 (illus)
      keypad, K-2, K-3 (table) (illus)
      lighting, battery, and function indicators and labels, K-4
      menus, K-7
      message, K-5
      multifunction keys, K-4
      pop-ups, K-5
      POS/PAGE set, K-7 to K-9 (illus)
   introduction, K-1 (table), K-2 (illus)
   manual initialization, K-16
   navigation, K-20
   new alert, creation of, K-19, K-20 (illus)
   new route, creation of, K-18, K-19 (illus)
   operating procedures, K-13 to K-15, K-16 (illus)

situational awareness, K-21 to K-24 (illus)
troubleshooting, K-24
waypoints, creation of, K-16 to K-17 (illus)
datum plane, 10-1, 3-3
DD Form (*see* forms)
dead reckoning, 11-14 to 11-16, 11-18, 12-5 to 12-7 (illus)
declination diagram, 6-7, 6-8 (illus)
   applications, 6-9 to 6-13 (illus)
   conversion, 6-8, 6-9 (illus)
   grid convergence, 6-8
   grid-magnetic angle, 6-8
   location, 6-8
defense advanced GPS receiver (*see* DAGR)
degree, 6-1
depression, 10-13 (illus), E-18 (illus) (*see also* terrain features)
depth perception (*see* stereovision)
desert, 13-1 (illus) (*see also* terrain, navigation in different types of)
   interpretation and analysis, 13-2
   navigation, 13-3
   regions, 13-1, 13-2 (table)
direction
   azimuths, 6-2 (illus)
   back, 6-3 (illus)
   grid, 6-3, 6-4 (illus)
   magnetic, 6-3
   base lines
      grid north, 6-2 (illus)
      magnetic north, 6-2 (illus)
      true north, 6-1, 6-2 (illus)
   declination diagram, 6-7, 6-8 (illus)
      applications, 6-9 to 6-13 (illus)
      conversion, 6-8, 6-9 (illus)
      grid convergence, 6-8
      grid-magnetic angle, 6-8 to 6-13 (illus)
      location, 6-8
   degree, 6-1
   grad, 6-1
   intersection, 6-13, 6-14 (illus)
   mil, 6-1

modified resection, 6-16, 6-17 (illus)
polar plot, 6-17 (illus)
protractor, 6-4 to 6-7 (illus)
resection, 6-15 (illus), 6-16 (illus)
distance
   graphic (bar) scales
      converting map distance to
         ground distance, 5-4, 5-5
         (illus), 5-7
      curved line, 5-5, 5-6 (illus)
      edge of paper exceeds, 5-5 to 5-8
         (illus)
      statute or nautical miles, 5-7
      time-distance scale, 5-8, 5-9
         (illus)
      using, 5-4 (illus)
   odometer, 5-10
   pace count, 5-10
   range estimation, 5-11 to 5-13 (illus)
      (table)
      100-meter unit-of-measure
         method, 5-11, 5-12 (illus)
      flash-to-bang method, 5-12
      proficiency of methods, 5-12
   representative fraction (RF), 5-1 to
      5-3 (illus)
   subtense, 5-11
DOD Single Line Item Requisition
   System Document (Manual), DD
   Form 1348, 2-2
DOD Single Line Item Requisition
   System Document (Mechanical),
   DD Form 1348M, 2-2
draw, 10-13, 10-14 (illus), E-16 (illus)
   (see also terrain features)
duties, navigator's, 12-1

elevation and relief
   contour intervals, 10-2 (illus) to 10-5
      (illus)
   definitions, 10-1
   depicting, methods of
      contour lines, 10-1, 10-2 (illus)
      form lines, 10-1
      hachures, 10-1
      layer tinting, 10-1

      shaded relief, 10-1
   foreign maps, H-1
   profiles, 10-19 to 10-22 (illus), 10-23
      (illus)
   slopes
      percentage of, 10-7 to 10-10
         (illus)
      types of, 10-5 to 10-7 (illus)
   terrain features
      cliff, 10-14, 10-15 (illus)
      cut, 10-15, 10-16 (illus)
      depression, 10-13 (illus)
      draw, 10-13, 10-14 (illus)
      fill, 10-15, 10-16 (illus)
      hill, 10-11 (illus)
      ridge, 10-12, 10-13 (illus)
      saddle, 10-11, 10-12 (illus)
      spur, 10-14 (illus)
      valley, 10-12 (illus)
EPLRS (Enhanced Position Location
   Reporting System), G-1 (see also
   navigation aids)
equipment (see navigation aids)

field sketches 2-7, A-1, A-2 (illus) (see
   also sketches)
fill, 10-15, 10-16 (illus), E-19 (illus) (see
   also terrain features)
foreign maps, 2-7, H-1
form lines, 10-1
forms
   DD Form 1348, (DOD Single Line
      Item Requisition System
      Document [Manual]), 2-2
   DD Form 1348M, (DOD Single Line
      Item Requisition System
      Document [Mechanical]), 2-2

geographic coordinates, 4-1 to 4-9
   (illus), 4-10 (table)
GPS (Global Positioning System), 9-12,
   G-2, I-1 (see also navigation aids)
grad, 6-1
graphic scales (see distance: graphic
   [bar] scales)

graphics, joint operations, 2-5 (illus), D-1
grid (*see also* grid systems)
   azimuth, 6-3, 6-4 (illus)
   coordinates, 4-14 to 4-17 (illus)
   coordination, locating a point using, 4-18 (illus) to 4-20 (illus)
   geographic coordinates, 4-1 to 4-9 (illus), 4-10 (table)
   military, 4-10
      Universal Polar Stereographic (UPS) grid, 4-12 (illus), 4-13
      Universal Transverse Mercator (UTM) grid, 4-11 (illus) to 4-13 (illus)
   north, 6-2 (illus)
   note, 3-2
   point designation, 8-15 (illus) to 8-18 (illus)
   reference box, 3-3, 4-23, 4-24 (illus)
   reference system, 4-1
grid systems (*see also* grid)
   British, 4-24
   GEOREF (world geographic reference system), 4-24 to 4-26 (illus)
   U.S. Army military grid reference system
      100,000-meter square, 4-13, 4-14 (illus)
      coordinates, 4-14 to 4-17 (illus)
      zone designation, 4-13, 4-14 (illus)
ground distance, 5-2 (illus)
G/VLLD (ground-vehicular laser locator designator), G-3 (*see also* navigation aids)

hachures, 10-1
handrails, 11-17
hasty profile, 10-22, 10-23 (illus)
hill, 10-11 (illus) (*see also* terrain features)

intersection, 6-13, 6-14 (illus)

joint operations graphics, 2-5 (illus), D-1
jungle, 13-6 (illus) (*see also* terrain, navigation in different types of)
   interpretation and analysis, 13-7
   navigation, 13-8, 13-9 (table)
   operations, 13-7

land navigation course, 14-2
latitude, 4-1 to 4-9 (illus), 4-10 (table)
layer tinting, 10-1
line map, 2-4
locations and map coordinates, protection of, 4-26
longitude, 4-1 to 4-9 (illus), 4-10 (table)

M2 compass, F-1 (illus) to F-3 (illus) (*see also* compasses)
   characteristics, F-2
   description, F-2
   use, F-3 to F-4 (illus)
magnetic north, 6-2 (illus)
maps
   accuracy, standards of, 2-7
   care, 2-3
   categories, 2-3, 2-4 (illus)
   colors used, 3-6
   coordinates and locations, protection of, 4-26
   definition, 2-1
   distance, 5-2 (illus)
   folding
      methods, B-1 (illus)
      practice cut, B-2
      protection, B-2 (illus)
   marginal information (*see main entry*)
   National Geospatial-Intelligence Agency (NGA), 2-1, 2-2
   orientation of, 11-1 to 11-5 (illus), 11-6 (illus)
   procurement, 2-2
   protection, B-2 (illus)
   purpose, 2-1

reading (*see also* grid systems *and* marginal information)
   grid coordination, locating a point using, 4-18 (illus) to 4-20 (illus)
   US Army military grid reference system, locating a point using, 4-20 to 4-23 (illus)
scale, 2-3, 2-4 (illus)
security, 2-2
substitutes, 2-6
   aerial photographs, 2-7 (*see also main entry*)
   atlases, 2-7
   city/utility maps, 2-7
   field sketches, 2-7
   foreign maps, 2-7, H-1
   geographic maps, 2-7
   tourist road maps, 2-7
symbols (*see main entry*)
types
   joint operations graphics, 2-5 (illus), D-1
   military city map, 2-6
   photomap, 2-5
   photomosaic, 2-6
   planimetric, 2-4
   special maps, 2-6
   terrain model, 2-6
   topographic, 2-5
marginal information, 3-4 (illus)
   adjoining sheets diagram, 3-2
   bar scales, 3-2
   classification, 3-5
   contour interval note, 3-2
   control note, 3-3
   conversion graph, 3-5
   coverage diagram, 3-5
   declination diagram, 3-2
   edition number, 3-1
   elevation guide, 3-2
   glossary, 3-5
   grid note, 3-2
   grid reference box, 3-3, 4-23, 4-24 (illus)
   horizontal datum note, 3-3
   index to boundaries, 3-2
   legend, 3-4
   preparation note, 3-3
   printing note, 3-3
   projection note, 3-2
   protractor scale, 3-5
   scale, 3-1
   series name, 3-1
   series number, 3-1
   sheet name, 3-1
   sheet number, 3-1
   special notes, 3-5
   spheroid note, 3-2
   stock number identification, 3-5
   unit imprint and symbol, 3-3
   user's notes, 3-5
   vertical datum note, 3-3
mean sea level, 10-1
measure, units of, and conversion factors, C-1 (table) to C-3 (table)
meridians, 4-1 to 4-9 (illus), 4-10 (table)
METT-TC, 11-10 to 11-12
mil, 6-1
military city map, 2-6
modified resection, 6-16, 6-17 (illus)
mountain, 13-4 (illus) (*see also* terrain, navigation in different types of)
   characteristics, 13-4
   climate, 13-5
   interpretation and analysis, 13-5
   major systems, 13-5 (table)
   minor systems, 13-5
   navigation, 13-5
mounted land navigation
   combination navigation, 12-7
   dead reckoning navigation, 12-5, 12-6 (illus) (*see also* dead reckoning)
   duties, navigator's, 12-1
   GPS (Global Positioning System), compatibility with, I-2 (*see also* GPS [global positioning system])
   movement, 12-1
   preparation, 12-2
   vehicle capabilities, 12-2, 12-3 (illus)

weather, 12-2
principles, 12-1
stabilized turret alignment
  navigation, 12-7
terrain association, 12-3 to 12-5
  (illus)

National Geospatial-Intelligence Agency
  (NGA), 2-1, 2-2, E-1
navigation aids (*see also* mounted land
  navigation *and* navigation methods
  and terrain, navigation in different
  types of)
  AN/PVS-5/5A, night vision goggles,
    G-1
  AN/PVS-7B/D, night vision goggles,
    G-1
  compasses (*see main entry*)
  DAGR (defense advanced GPS
    receiver), K-20
  EPLRS (enhanced position location
    reporting system), G-1
  GPS (global positioning system),
    G-2 (*see also main entry*)
  G/VLLD (ground-vehicular laser
    locator designator), G-3
  PADS (position and azimuth
    determining system), G-2
  PLGR (precision lightweight global
    positioning system receiver) (*see
    main entry*)
  QRMP (quick response multicolor
    printer), G-3
navigation course, 14-2 (*see also* terrain,
  navigation in different types of)
navigation methods (*see also* mounted
  land navigation *and* navigation aids
  and terrain, navigation in different
  types of)
  dead reckoning, 11-14 to 11-16,
    12-5 to 12-7 (illus)
  dead reckoning and terrain
    association, combination of,
    11-18

field-expedient methods
  shadow-tip, 9-7 (illus)
  star, 9-9 to 9-12 (illus)
  watch, 9-8, 9-9 (illus)
night, 11-18
terrain association, 11-16 to 11-18
  (illus)
night vision goggles (*see* navigation
  aids)

OCOKA, 11-9
odometer, 5-10
orienteering
  civilian, E-20
  clue description card, E-9 (illus)
  control markers, E-6 (illus)
  control point guidelines, E-10
  course, E-1
    cross-country, E-3 (illus)
    line, E-2 (illus)
    route, E-2
    score, E-4 (illus)
  description, E-1
  equipment, E-5
  event card, E-8 (illus)
  history, E-1
  map symbols, E-11 (illus) to E-19
    (illus)
  officials, E-5
  prizes, E-10
  recorder's sheets, E-6, E-7 (illus)
  safety, E-10
  scoring, E-9
  start/finish area
    assembly area, E-5
    clue description card, E-9 (illus)
    equipment, E-5
    event card, E-8 (illus)
    master map area, E-5
    prizes, E-10
    recorder's sheets, E-6, E-7 (illus)
    results board, E-8 (illus)
    scoring, E-9
    start, E-5

techniques
  attack points, E-19
  handrails, E-19, E-20 (illus)
  pacing, E-19 (*see also* pace count)
  thumbing, E-19
overlays
  aerial photograph (*see also* main entry)
    marginal information, recording, 7-3, 7-4 (illus)
    orienting, 7-3
  map, 7-1 (illus)
    marginal information, recording, 7-2, 7-3 (illus)
    new detail, plotting, 7-2
    orienting, 7-1
  purpose, 7-1

pace count, 5-10, E-19
PADS (Position and Azimuth Determining System), G-2 (*see also* navigation aids)
parallels, 4-1 to 4-9 (illus), 4-10 (table)
photographs (*see* aerial photographs)
photomap, 2-5
photomosaic, 2-6
planimetric map, 2-4
PLGR (precision lightweight Global Positioning System receiver) (*see also* navigational aids)
  capabilities, J-1
  characteristics, J-2 (illus)
  navigation, J-7 to J-9 (illus)
  operation, concept of, J-1
  setup and control, J-3 (illus) to J-5 (illus)
  waypoint operations, J-6 (illus), J-7 (illus)
polar plot, 6-17 (illus)
prime meridians, 4-1 to 4-9 (illus), 4-10 (table)
profiles, 10-19 to 10-22 (illus), 10-23 (illus)
protractor, 6-4 to 6-7 (illus)

QRMP (quick response multicolor printer), G-3 (*see also* navigation aids)

range estimation, 5-11 to 5-13 (illus) (table)
  100-meter unit-of-measure method, 5-11, 5-12 (illus)
  flash-to-bang method, 5-12
  proficiency of methods, 5-12
relief (*see* elevation and relief)
representative fraction (RF), 5-1 to 5-3 (illus)
resection, 6-15 (illus), 6-16 (illus)
RF (representative fraction), 5-1 to 5-3 (illus)
ridge, 10-12, 10-13 (illus), E-15 (illus) (*see also* terrain features)
ridgeline, 10-10, 10-11 (illus)
ridgelining, 10-18, 10-19 (illus)

saddle, 10-11, 10-12 (illus), E-15 (illus) (*see also* terrain features)
safety, 1-2
scale, 2-3, 2-4 (illus), H-1 (*see also* distance)
scale determination, aerial photograph, 8-8
  comparison method, 8-9 (illus)
  focal length-flight altitude method, 8-10 (illus)
security, 2-2
shaded relief, 10-1
sketches
  area, A-1, A-2 (illus)
  field, 2-7, A-1, A-2 (illus)
  purpose, A-1
  road, A-1, A-2 (illus)
slopes
  percentage, 10-7 to 10-10 (illus)
  types
    concave, 10-6, 10-7 (illus)
    convex, 10-7 (illus)
    gentle, 10-5, 10-6 (illus)
    steep, 10-6 (illus)

SOSES (shape, orientation, size, elevation, and slope), 10-18
special maps, 2-6
spot elevations, 10-5
spur, 10-14 (illus), E-16 (illus) (*see also* terrain features)
stars, 9-9 to 9-12 (illus)
steering marks, 11-14 to 11-16
stereovision, 8-19 to 8-23 (illus)
    mirror stereoscope, 8-22 (illus)
    pocket stereoscope, 8-21, 8-22 (illus)
streamlining, 10-19 (illus)
subtense, 5-11
sustainment (*see* unit sustainment)
symbols, map
    military, 3-6
    topographic, 3-5, E-11 (illus) to E-19 (illus)

terrain association
    locations, 11-6
    mounted, 12-3 to 12-5 (illus)
    movement and route selection, 11-12 to 11-14
    navigation methods
        dead reckoning, 11-14 to 11-16, 12-5, 12-6 (illus)
        dead reckoning and terrain association, combination of, 11-18
        terrain association, 11-16 to 11-18 (illus)
    night navigation, 11-18
    orientation of the map
        compass, using a, 11-1 to 11-3 (illus), 11-4 (illus)
        field-expedient methods, using, 11-5, 11-6 (illus)
        terrain association, using, 11-4, 11-5 (illus)
    tactical considerations
        METT-TC, 11-10 to 11-12
        OCOKA, 11-9
    usage, 11-6 to 11-9 (illus)

terrain features
    interpretation of, 10-16, 10-17 (illus)
        contour lines, 10-17
        ridgelining, 10-18, 10-19 (illus)
        SOSES (shape, orientation, size, elevation, and slope), 10-18
        streamlining, 10-19 (illus)
    types
        cliff, 10-14, 10-15 (illus)
        cut, 10-15, 10-16 (illus), E-19 (illus)
        depression, 10-13 (illus), E-18 (illus)
        draw, 10-13, 10-14 (illus), E-16 (illus)
        fill, 10-15, 10-16 (illus), E-19 (illus)
        hill, 10-11 (illus)
        ridge, 10-12, 10-13 (illus), E-15 (illus)
        saddle, 10-11, 10-12 (illus), E-15 (illus)
        spur, 10-14 (illus), E-16 (illus)
        valley, 10-12 (illus), E-17 (illus)
terrain model, 2-6
terrain, navigation in different types of
    arctic
        interpretation and analysis, 13-9
        navigation, 13-10
        operations, 13-9
    desert, 13-1 (illus)
        interpretation and analysis, 13-2
        navigation, 13-3
        regions, 13-1, 13-2 (table)
    jungle, 13-6 (illus)
        interpretation and analysis, 13-7
        navigation, 13-8, 13-9 (table)
        operations, 13-7
    mountain, 13-4 (illus)
        characteristics, 13-4
        climate, 13-5
        interpretation and analysis, 13-5
        major systems, 13-5 (table)
        minor systems, 13-5
        navigation, 13-5

urban areas
    interpretation and analysis, 13-10
    navigation, 13-11
time-distance scale, 5-8, 5-9 (illus)
topographic map, 2-5, 3-4 (illus)
training strategy
    Armywide implementation, 1-2
    building-block approach, 1-1
    safety, 1-2
train-the-trainer program, 14-2
true north, 6-1, 6-2 (illus)

unit sustainment
    GPS (global positioning system), compatibility with, I-2 (*see also* GPS [Global Positioning System])
    land navigation course, 14-2
    program, 14-1
    train-the-trainer program, 14-2
universal polar stereographic (UPS) grid, 4-12 (illus), 4-13
universal transverse Mercator (UTM) grid, 4-11 (illus) to 4-13 (illus)
UPS (universal polar stereographic) grid, 4-12 (illus), 4-13
urban terrain (*see also* terrain, navigation in different types of)
    interpretation and analysis, 13-10
    navigation, 13-11
UTM (universal transverse Mercator) grid, 4-11 (illus) to 4-13 (illus)

valley, 10-12 (illus), E-17 (illus) (*see also* terrain features)

This Page intentionally left blank.

FM 3-25.26
18 January 2005

ler of the Secretary of the Army:

PETER J. SCHOOMAKER
General, United States Army
Chief of Staff

l:

ANDRA R. RILEY
istrative Assistant to the
cretary of the Army
0501006

IBUTION:

Army, Army National Guard, and US Army Reserve: To be distributed in lance with initial distribution number 110166, requirements for FM 3-25.26.

This Page intentionally left blank.

This Page intentionally left blank.

FOUO

PIN: 079'

www.ingramcontent.com/pod-product-compliance
Lightning Source LLC
Chambersburg PA
CBHW050049230526
45470CB00004B/1463